Asia-Pacific STEM Teaching Practices

Ying-Shao Hsu · Yi-Fen Yeh
Editors

Asia-Pacific STEM Teaching Practices

From Theoretical Frameworks to Practices

 Springer

Editors
Ying-Shao Hsu
National Taiwan Normal University
Taipei, Taiwan

Yi-Fen Yeh
National Taiwan Normal University
Taipei, Taiwan

ISBN 978-981-15-0770-0   ISBN 978-981-15-0768-7   (eBook)
https://doi.org/10.1007/978-981-15-0768-7

This Springer imprint is published by the registered company Springer Nature Singapore Pte Ltd.
The registered company address is: 152 Beach Road, #21-01/04 Gateway East, Singapore 189721, Singapore

Contents

Chapter 1
Opportunities and Challenges of STEM Education

Ying-Shao Hsu and Su-Chi Fang

1.1 Introduction

STEM (science, technology, engineering, and mathematics) education has emerged in educational reforms in the past years worldwide. STEM education can be viewed as an instructional approach in which the contents of the STEM disciplines are addressed as a cluster of individual science, technology, engineering, and mathematics ideas (multidisciplinary) or as integrated ideas (interdisciplinary and transdisciplinary) into the engineering design process and open-ended inquiry into real-world issues and problems. It is believed that STEM education has the potential to motivate students to study and engage in future STEM fields. More importantly, STEM education can prepare students to tackle multidisciplinary and increasingly complex problems in a global society.

In recent years, the ubiquitous calls for STEM education have encouraged educators and policymakers to promote STEM practices in classrooms for the next generation of citizens. However, there seems to be no consensus on the definition of STEM. Some researchers define STEM as a broad area encompassing many disciplines and epistemological practices (Lamb, Akmal, & Petrie, 2015). Others regard it as applying transdisciplinary knowledge and skills in solving real-world problems (Breiner, Harkness, Johnson, & Koehler, 2012). The definitions do not consistently address whether STEM is an umbrella label for a cluster of scientific/mathematical-oriented economic opportunities and careers or an integrated discipline with potential synergies not found in the separate disciplines. Shernoff, Sinha, Bressler, and Ginsburg (2017) suggested that STEM education is to create student-centered learning environments in which students construct evidence-based explanations of phenomena

Ying-Shao Hsu is a visiting professor at University of Johannesburg, South Africa.

Y.-S. Hsu (✉) · S.-C. Fang
Graduate Institute of Science Education, National Taiwan Normal University, Taipei City, Taiwan
e-mail: yshsu@ntnu.edu.tw

© Springer Nature Singapore Pte Ltd. 2019
Y.-S. Hsu and Y.-F. Yeh (eds.), *Asia-Pacific STEM Teaching Practices*,
https://doi.org/10.1007/978-981-15-0768-7_1

and then generate solutions to real-world problems through applying science, technology, engineering, mathematics, and associated practices. While students construct explanations about the world and seek solutions for problems, they are given opportunities to develop critical thinking, collaborative approaches, and communicative skills. Furthermore, such student-centered learning environments allow more space for their creative and innovative skills development.

The number of STEM education studies reported in science education journals and other publications has grown in parallel with the calls for STEM education from national policies and promotions by various layers of the educational system. This chapter is intended to present a vision of STEM education based on the review of and reflection on these studies. From these insights into STEM research, we identify and address the challenges faced by teachers and educators. We also point out possibilities by proposing a comprehensive, multifaceted, and coherent STEM education vision for teachers and teacher educators.

1.2 Conceptualizing STEM Education in the Field of Science Education

We collected and reviewed the studies relevant to integrated STEM education in the five key science education journals: *Journal of Research in Science Teaching, Science Education, International Journal of Science Education, International Journal of Science and Mathematics Education*, and *Research in Science Education*. The analysis attempted to understand how STEM education is conceptualized in the field of science education by considering these studies' research foci, goals of the STEM programs, and how STEM education was integrated, designed, and implemented.

The first database ($N = 165$ articles) was formed using the keyword *STEM* to search in the five journals from 1995 through 2018. After eliminating research studies that were not related to designing, implementing, and evaluating integrated STEM curricula, 26 articles (identified by * in References) were retained for the final content review. The following sections of this chapter report on the trends revealed from the review regarding the foci of STEM education research, the intended goals of these STEM programs, the nature of STEM integration, and implementation supports.

1.2.1 The Foci of STEM Education Research

Two major research themes were found in the reviewed studies. The first research theme was concerned with the effects of integrated STEM curricula on student learning and achievement; the second research theme focused on how integrated STEM curricula might impact students' engagement, interests, identity, and career choices toward STEM.

1.2.1.1 Learning and Achievement

The review adopted different levels of achievement assessment (Ruiz-Primo, Shavelson, Hamilton, & Klein, 2002) to evaluate the effects of integrated STEM curricula on students' learning outcomes and to determine the factors that influenced the STEM learning experiences.

Some studies used the assessments at the proximal level (Ruiz-Primo et al., 2002) such as unit tests to evaluate student content knowledge gains on specific scientific topics after experiencing a short-term STEM curriculum. Korur, Efe, Erdogan, and Tunç (2017) showed that using a scaffolded, design-based approach (to create a toy crane) increased middle school students' conceptual understanding of simple machines and their creative attitude. Park, Park, and Bates (2018) reported that engineering design practices in a STEM activity (create a clay boat) could enhance young children's understandings about the concept of volume. King and English (2016) explored how students applied STEM concepts in an optical engineering design process rather than simply focusing on content knowledge. Chien and Chu (2017) analyzed students' worksheets, design drafts, and group discussions after a CO_2-car engineering design unit in order to understand how they applied scientific knowledge in the design process and how their creativity might be improved.

Another group of researchers investigated to what extent long-term STEM experiences might impact students' knowledge gains. Guzey, Ring-Whalen, Harwell, and Peralta (2019) reported that a 3-year, design-focused, life science curriculum resulted in higher student science and engineering learning gains. Means, Wang, Young, Peters, and Lynch (2016) compared the achievement of northern California students from inclusive STEM high schools (ISHSs) and non-STEM comparison high schools in terms of grade point average (GPA) and American College Testing scores. Their findings indicated that attending ISHSs had positive impacts on GPA scores. Similarly, their second study also showed positive relationships between ISHSs attendance and GPA scores in two samples from different places: North Carolina and Texas (Means, Wang, Young et al., 2017).

Apart from examining the effects of STEM experiences on the overall student achievement, several studies explored how different (e.g., gender, ethnicity, achievement level) groups of students responded to STEM curricula differently and led to diverse learning outcomes. Han, Capraro, and Capraro (2015) employed a state-wide accountability assessment to measure student mathematics achievement; they found that, compared to middle and high achievers, STEM project-based learning (PBL) activities benefited low-performing students the most. Dickerson, Eckhoff, Stewart, Chappell, and Hathcock (2014) pointed out that a pullout STEM program was able to close the achievement gap in terms of standardized test scores between Black and White students. It was also reported elsewhere that grouping students with diverse academic preparedness was highly associated with positive learning outcomes and benefited less-prepared students (Micari, Van Winkle, & Pazos, 2016).

The overall findings indicated that STEM programs were promising for enhancing students' science learning achievement and especially helpful for underrepresented populations in STEM fields. Nevertheless, as the studies did not provide critical information about long-term STEM programs—such as to what extent the goals of the STEM programs are aligned with state-level examinations and how those integrated STEM curricula were designed, implemented, and evaluated in a school-wide manner—little can be concluded from the findings about what ways the STEM programs supported student learning.

1.2.1.2 Engagement, Interest, Identity, and Careers Toward STEM

The impact of integrated STEM curricula on students' affective learning outcomes, self-related constructs, and career options was the other major theme being studied in this review. Generally, long-term STEM experiences appear to raise high school and university students' interest in science and technology and encourage them to pursue a STEM career. Two studies by the Means' research group revealed that the 3-year ISHSs experiences raised the tendency that a student will complete calculus and chemistry in high school, participate in STEM-related extracurricular activities, study for higher degrees, and pursue science careers (Means, Wang, Wei et al., 2017; Means, Wang, Young et al., 2016). Moreover, female and underrepresented minority students in ISHSs were more likely to undertake STEM coursework and express their career interest in scientific areas. Similar positive effects on STEM career aspirations were found for students who participated in high school STEM summer programs: the National Science Foundation's STEM Talent Expansion Programs (Kitchen, Sonnert, & Sadler, 2018) and an 8-week, college-level, environmental science program (Romine & Sadler, 2016). Relatively shorter STEM programs, such as a 5-day pullout primary STEM program (Dickerson et al., 2014), and Science Olympiad competition experiences (Sahin, Gulacar, & Stuessy, 2015), were found to be helpful for boosting student interest in learning science and STEM careers.

It is worth noting that the effects of STEM programs on students' science achievement, interest, self-efficacy, and career orientation are intertwined. Lamb et al. (2015) used a survey and state-created tests to explore the interplay among student science content, cognitive, and affective outcomes. They concluded that affective traits (i.e., science interest and self-efficacy) had positive, direct, and recursive effects on science content scores, whereas cognitive attributes (i.e., spatial visualization and mental rotation) had positive, direct, and linear effects on science content scores. Adedokun, Bessenbacher, Parker, Kirkham, and Burgess (2013) studied students participating in a long-term, faculty-mentored interdisciplinary STEM research program and found that their research skills and research self-efficacy were good predictors for their aspirations for research careers.

Despite the many promising outcomes, there were research findings related to secondary students that showed conflicting results. Guzey et al. (2019) indicated that, although the middle school students had significant learning gains after a 3-year design-focused course, their interest in science and engineering did not change. Schütte and Köller (2015) found that a 2-year intervention program implemented as an elective course did not increase secondary students' science motivation to pursue a science career. One possible explanation is a ceiling effect in that the students in the elective course were already considerably more motivated than their counterparts.

The studies interested in exploring the influence of STEM experiences on the formation of students' science or STEM identity tended to use a mentorship or apprenticeship approach implemented in informal learning environments. Their findings showed that providing role models and authentic research activities could improve secondary students' STEM identities (Burgin, McConnell, & Flowers, 2015; Hughes, Nzekwe, & Molyneaux, 2013), and this effect was particularly found for girls (Hughes et al., 2013; Todd & Zvoch, 2017). Krogh and Andersen (2013) investigated students' trajectories toward science majors and careers and found that four factors (i.e., general identity process orientation, personal values, subject self-concept, and subject interest) interact with each other and set students' future education and career search directions.

1.2.2 STEM Program Goals

The goals of a STEM program are important as they guide instructional design and drive intended educational interventions. We found in this review that some of the research did not explicitly identify the goals of the STEM program (e.g., Guzey et al., 2019; Krogh & Andersen, 2013), nor did it explain how the goals are interpreted and transformed into instructional design and approaches. It should be noted that, without considering the goals, the interpretation of the findings in terms of the impact of the STEM program on certain aspects of student learning performance might be problematic and misleading.

Two frequently cited goals of a STEM program are to raise student interest and engagement in the STEM subjects and to increase and develop a STEM-capable workforce. Schütte and Köller (2015) introduced a program that was implemented as an elective course aiming to increase and sustain student interest in science and technology. The ISHSs (Means Wang, Wei et al., 2017; Means, Wang, Young et al., 2016) and the summer science camp (Todd & Zvoch, 2017) emphasized student interest in STEM fields. These intensive STEM programs and informal learning environments were intended to engage more students from traditionally under-represented groups (e.g., girls or students from low socioeconomic status backgrounds) into STEM learning areas and prepare them for college and STEM careers. Other studies explored different approaches (e.g., providing scientific literature-based learning experiences, engaging in university laboratories, and exposure to authentic experiences about STEM careers) to encourage high school or undergraduate students to pursue STEM careers (Adedokun et al., 2013; Burgin et al., 2015; Hughes et al., 2013; Kitchen et al., 2018; Romine & Sadler, 2016).

The other common goal set by STEM programs is relevant to the promotion of student learning of STEM subject content. In addition to enhancing student understanding of fundamental concepts, STEM programs emphasize the connections across

STEM subjects and the applications of STEM concepts. Two approaches were frequently used: having students solve complex, authentic real-world (Lamb et al., 2015) or advanced, conceptually rich (Micari & Light, 2009; Micari et al., 2016) problems and engaging students in engineering design tasks (Chien & Chu, 2017; King & English, 2016; Park et al., 2018; Schnittka, Evans, Won, & Drape, 2016). Some teacher professional development (PD) programs related to STEM adopted similar foci and approaches as their program goals (Brown & Bogiages, 2019). A few STEM programs focused on rather specific goals, for example, to facilitate the learning of target scientific or mathematical concepts in STEM contexts (Korur et al., 2017; Prieto & Dugar, 2017) or to raise and heighten student awareness of sustainability challenges (Sahin et al., 2015).

1.2.3   Integrated STEM Education

Integrated (interdisciplinary or transdisciplinary) curricula is one significant feature of STEM education. This section analyzes and reports the type of connections presented in the STEM curricula and the instructional approaches used to illustrate the nature of integration implemented in and the challenges flowing from the research studies reviewed.

1.2.3.1   Nature of STEM Integration: Type of Connections, Disciplines Emphasized, and Instructional Approaches

K–12
In general, the research studies revealed diverse ways to connect STEM subjects, placed emphases on different disciplines, and employed multiple instructional approaches. Lamb et al. (2015) developed a long-term, whole-school STEM program that explicitly embedded STEM concepts into and across the curriculum using design-based, project-based, problem-based, and inquiry learning approaches. Apart from school-based experiences, the program also created opportunities to engage students in hands-on laboratory and experimental learning with STEM professionals. Han, Capraro et al. (2015) created a STEM PBL curriculum that included all of the STEM disciplines but put emphasis on science and mathematics content.

A few studies used engineering design as a learning context to integrate STEM subject content (Chien & Chu, 2017; King & English, 2016). Chien and Chu (2017) engaged students in CO_2-car engineering design activities using handmade, laser-cut, and 3D-printing skills. King and English (2016) designed engineering-based activities (i.e., aerospace, civil, and materials engineering) with the intention of complementing and building on existing mathematics, science, and technology curricula. Guzey et al. (2019) utilized the *Next Generation Science Standards* (National

Research Council, 2013) science and engineering practices to construct engineering design challenges focused on life science concepts rather than all of the STEM subjects.

Other studies adopted various transdisciplinary topics, such as forensics (Todd & Zvoch, 2017) and weather and human impact on the environment (Dickerson et al., 2014), or scenarios, such as Save the Penguins (Schnittka et al., 2016), to create integrated STEM learning environments for classrooms. The use of transdisciplinary topics was also extended to informal educational settings. Burgin et al. (2015) and Hughes et al. (2013) used research apprenticeship programs with practicing scientists to provide students genuine laboratory experiences and to engage them in solving real-world environmental or engineering problems. Sahin et al. (2015) used student participation in various competitions on international sustainable world energy, engineering, and environmental projects as STEM learning experiences.

Among the reviewed studies focused on K–12 levels, about one-third of the studies (6/17) did not specify how different disciplines were integrated or what type of instructional approach was adopted in the curriculum design. Two large-scale studies (Means, Wang, Wai et al., 2017; Means, Wang, Young et al., 2016) involved participants from more than ten ISHSs. Although four key ISHS features were presented, the nature and the scope of the STEM curriculum implemented in these ISHSs were not specified. In some cases, the curriculum seemed to be individualized, providing students the freedom to develop individual projects they found interesting (Schütte & Köller, 2015) or to select a university mentorship program for aspiring scientists (Krogh & Andersen, 2013). There were also cases that involved students in a science or engineering design-based process without explaining how and to what extent different disciplines were integrated (Korur et al., 2017; Park et al., 2018).

College and University Level

There were five studies (5/26) that explored STEM learning at the postsecondary level, where most of these STEM programs were set in informal learning environments. The STEM curricula involved in these programs were not structured and heavily depended on individual situations. Both the Undergraduate Research Experiences Program (Adedokun et al., 2013) and the STEM Talent Expansion Program (Kitchen et al., 2018) involved students in interdisciplinary STEM research laboratories and provided science and mathematics education beyond the high school level. Similarly, the Gateway Science Workshop Program (Micari & Light, 2009; Micari et al., 2016) engaged undergraduate participants in conceptual-rich problems created by faculty members in five STEM disciplines: biology, physics, chemistry, mathematics, and engineering. Students were required to apply various conceptual knowledge in real-world situations through the peer-led, small-group, problem-solving processes. Romine and Sadler (2016) designed literature interventions that supported students to read and analyze academic articles on environmental science through the scientific practice approach.

Teacher Education

Two studies explored preservice teacher education and PD programs. Brown and Bogiages (2019) appeared to stress science, mathematics, and engineering instructional methods separately instead of an integrated approach. Carrier, Whitehead,

Walkowiak, Luginbuhl, and Thomson (2017) held PD workshops aiming to engage teachers in integrated STEM tasks and to exemplify connections between mathematics practices and science and engineering practices.

1.2.4 Implementation Supports

Successful development and implementation of integrated STEM curricula must support classroom teachers in overcoming the traditions of separated school disciplines. It is critical to provide educators opportunities to experience and practice subject-matter integration and to improve their expanded pedagogical practices. Among the 22 studies at the K–12 and college levels, nearly half (9/22) did not mention if the STEM program provided support to teachers (Adedokun et al., 2013; Hughes et al., 2013; Kitchen et al., 2018; Korur et al., 2017; Means, Wang, Wei et al., 2017; Means, Wang, Young et al., 2016; Romine & Sadler, 2016; Sahin et al., 2015; Schütte & Köller, 2015).

However, three different types of implementation support were identified in the remaining studies. Collaborating with universities was one common approach. University faculty members and teachers formed a partnership to design and implement STEM curricula in some studies (Chien & Chu, 2017; Dickerson et al., 2014; King & English, 2016; Krogh & Andersen, 2013; Tippett & Milford, 2017). Others involved educational researchers or STEM center staff conducting workshops to improve the instructional practices of teachers, group leaders, or peer leaders (Han, Capraro et al., 2015; Micari & Light, 2009; Micari et al., 2016; Park et al., 2018; Schnittka et al., 2016; Todd & Zvoch, 2017) or to facilitate their instructional design abilities to integrate engineering into science teaching (Guzey et al., 2019). Lamb et al. (2015) adopted a team-teaching model that recruited a STEM coordinator to engage teachers in PD and curriculum development. This whole-school STEM program exploited varied resources (e.g., university laboratories, science centers, and museums) to enrich students' STEM learning experiences. Burgin et al. (2015) employed a mentorship approach that engaged professional scientists or engineers and their laboratory research groups to support student learning.

1.2.5 Summary of the Review

The reviewed research on integrated STEM in science education appeared to focus on two themes: the impact of STEM curricula on students' academic achievements and the impact on students' interests in learning and perusing careers in STEM fields. Though most of the research findings showed promising effects, there is limited diversity in the influence of integrated STEM. For instance, little is known about how and to what extent integrated STEM learning experiences may foster student creativity, support the development of higher order thinking skills, or impact their epistemological beliefs and views about science learning. Because the nature of STEM integration in terms of the type of connections, the disciplines being emphasized, and highly diverse instructional approaches, little can be concluded from the research about the effectiveness of various teaching practices (e.g., instructional design, teaching strategies, etc.). Moreover, the review indicated that only a few studies looked into the issues about the preparation of STEM teachers—initial teacher education and PD programs—on integrated STEM. More research is needed to help preservice and in-service teachers develop expertise for teaching integrated STEM. Based on the gaps identified in this review, the next section will identify the challenges encountered in developing and implementing integrated STEM.

1.3 Challenges

The current STEM education debates on integration across disciplines and teacher preparation have identified some barriers to the advancement of STEM education as an interdisciplinary approach in K–12 including the shortage of qualified teachers, lack of PD for teachers, poor motivation of students, weak connection with individual learners, little support from the school system, poor content without integration across disciplines, lack of quality assessments, poor facilities, and lack of hands-on experiences for students (Ejiwale, 2013). Furthermore, opinions on how STEM education should be enacted vary across school contexts, curricula, and educational policies (English, 2017). These challenges led us to consider the following important issues in STEM education:

- perspectives on the nature of STEM education,
- approaches to STEM integration,
- pedagogies for integrated STEM, and
- preparation of teachers' STEM pedagogical content knowledge (PCK).

The nature of STEM education varies in scope and specificity (English, 2017). STEM education might be referenced as a collective of four disciplines but where only one discipline is emphasized (e.g., Chien & Chu, 2017) or four disciplines are presumed to be separate but equal (e.g., King & English, 2016). Some researchers and educators highlight one or two specific disciplines within the STEM space (Shaughnessy, 2013); others define STEM education as a holistic integration of the four disciplines (National Academy of Engineering & National Research Council, 2014). Within these various definitions, STEM integration appears to be increasingly emphasized for the interdisciplinary solution of real-world problems (Bryan, Moore, Johnson, & Roehrig, 2016; English, 2016). Therefore, it is necessary to identify the fundamental content and processes of the respective disciplines for students' core competencies in STEM learning (English, 2017). The core STEM competencies for students can be identified as essential abilities for solving problems (Shaughnessy,

2013). Shaughnessy's definition refers to the procedures of solving problems while incorporating teamwork, application of concepts and skills in science and mathematics, design methodology of engineering, and utilization of appropriate technology.

One big challenge for STEM educators and researchers is how multiple disciplines should be integrated into a curriculum as a whole and implemented without losing disciplinary integrity (English, 2017). Numerous frameworks for implementing STEM integration have been proposed (e.g., Bryan et al., 2016; Vasquez, Sneider, & Comer, 2013). This can be seen as a continuum from multidisciplinary, interdisciplinary to transdisciplinary approaches not only involving concepts and skills in each discipline but also knowledge and skills that can be learned from two or more disciplines and applied to real-world problems (English, 2017; King & English, 2016). The activities associated with integrated STEM problems (e.g., constructing an earthquake-proof building, applying engineering techniques to slow coastline erosion, constructing a bridge) are essential for consolidating and shaping transdisciplinary learning experiences and facilitating learners' STEM-related core competencies—conceptual understanding, attitude, and practices.

Another challenge for successful STEM education is how to strengthen teachers' beliefs and nurture their knowledge of integrated STEM. Possible considerations include how to

- provide opportunities for teachers to construct perspectives and a vision of STEM education;
- enhance teaching skills of engaging and motivating learners to solve integrated STEM problems;
- develop appropriate pedagogical tools (e.g., analogies, models);
- provide sustained support to STEM education communities.

The educational system needs to create an atmosphere for supporting the implementation of STEM education in schools, to hold PD workshops, to seek resources from government and industry, and to help establish communities across various disciplines.

Teachers struggle to become facilitators for STEM education without being equipped with STEM PCK. Enhancing teachers' STEM PCK—defined as teachers' knowledge of students' thinking about STEM topics, knowledge of curriculum, teaching and assessment strategies, and knowledge of real-world STEM-related issues and their complexity (Allen, Webb, & Matthews, 2016)—can effectively promote integrated STEM education. Teachers' STEM PCK needs to be cultivated through authentic practices in PD workshops and classrooms.

1.4 Design Framework of STEM Curricula

STEM education aims to help students develop their problem-solving competencies and computational thinking (e.g., English, 2017) through the use of science, technology, engineering, and mathematics and their associated practices to solve a real-world problem (e.g., Schnittka et al., 2016). During STEM learning, students act as an autonomous learner to self-regulate their problem-solving process and transfer their knowledge and skills to new real-world contexts. Therefore, a pressing issue is: how do teachers prepare a student-centered learning environment that helps students develop STEM literacy, specifically, their inquiry, critical thinking, creative, collaborative, and communication skills. We propose the following 5-step framework for developing STEM curricula:

1. Identify core competencies.
2. Select a real-world context or problem.
3. Prepare supporting resources and tools.
4. Design a series of activities to engage students.
5. Develop an evaluation rubric for assessing the selected core competencies.

This framework ensures curriculum experiences in which students gain the necessary support from teachers and resources. For example, students need to select proper tools for a certain situation: students need to apply a statistical tool to estimate the probability of an event or occasion for their decision-making or evaluating the level of fit for a model. In addition, students are required to apply computational thinking to analyze the complexity of the real-world situation through identifying components within it and the relationship between the components and, then, to develop the several alternative solutions of the real-world situation through investigation, evidence-based explanation, and evaluation of outcomes.

The following example illustrates the process of using the 5-step design framework to develop an integrated STEM learning module. We used the idea of "building functional models: designing an elbow" (Penner, Giles, Lehrer, & Schauble, 1997) to develop an integrated STEM learning module as an example for illustrating the framework (Fig. 1.1). In the first step, we select three core competencies for this integrated STEM lesson including analogical reasoning, quantitative thinking, and reflective ability. In the second step, a real-world problem—how to design an artificial elbow for injured patients—is chosen to promote these three core competencies among the students. The goals of this lesson are to support students to learn how to build a model that can work like a human elbow and how to test the quality of their model for the better solution. In the third step, teachers provide resources (e.g., websites, materials for building the model) and tools (e.g., software of drawing their model, mathematics software for calculation, and 3D printers). In the fourth step, teachers design a series of activities to engage and motivate students to complete their tasks and apply specific competencies (indicated by colored dots in Fig. 1.1), including

- identifying the problem and breaking down the problem into a few tasks for a possible solution;
- finding an analogy for the model and drawing a design diagram of the model (analogical reasoning);
- making a plan with material selections and calculations (quantitative thinking);
- building an artificial elbow;

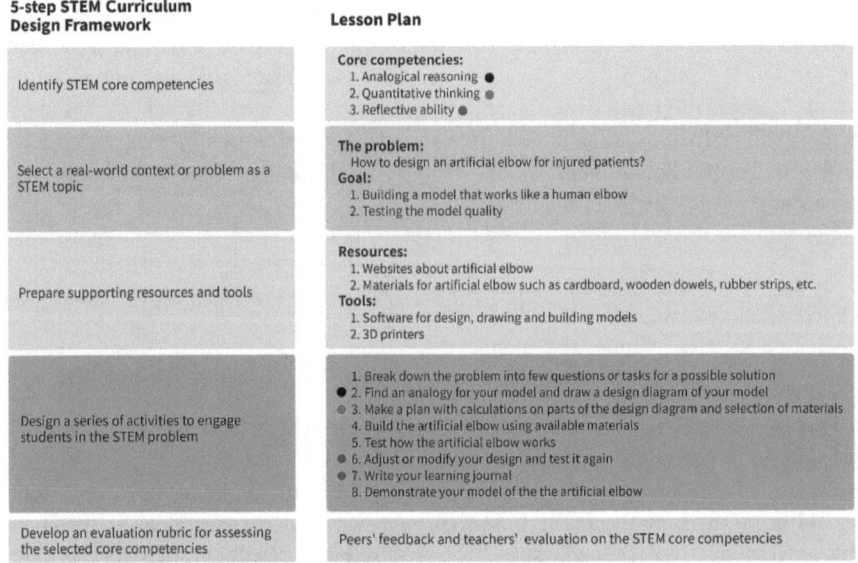

Fig. 1.1 A lesson plan of a 5-step STEM curriculum design framework

- testing how the artificial elbow works;
- adjusting or modifying the design and testing it again (reflective ability);
- writing reflections in a journal on STEM learning (reflective ability);
- demonstrating the artifact of the model.

When students attempt to bring their previous experiences to design an elbow, they need to apply analogical reasoning to think about the function of the elbow as that of a door hinge. The design diagram requires students to select the materials and calculate the strength of each part of the model in order to make the model work well. Then, students are required to find a way to test the model and make judgments about the model quality. Finally, they write a learning journal and demonstrate the model to their peers. Teachers guide students to look back at their design and facilitate peer discussions and feedback for reflective learning. In the fifth step, teachers provide comments across different model designs and summarize some possible issues for students' further thoughts about this integrated STEM lesson. This 5-step lesson plan should be conducted under teamwork contexts to cultivate (i.e., induce, develop, reinforce) students' three STEM-related core competencies.

1.5 Professional Preparation for STEM Education

Integrated STEM lessons need to engage students in real-world situations and nurture their interest in developing the necessary competencies and practical skills of STEM professions. Teachers' preparation is the key to unlocking the gate to STEM literacy. Therefore, teachers need help to frame and reframe their practical knowledge for STEM teaching, which is central to STEM teacher education and professional development.

In order to enact integrated STEM education, teachers need to change their professional practice, improve their understanding of STEM content, and increase their STEM PCK (Saxton et al., 2014). Successful STEM education requires teachers to adjust their instruction in productive ways through developing the necessary knowledge to recognize and support their students' STEM-related conceptual development, inquiry and thinking processes, and understanding of disciplinary connections and real-world complexity. (More discussions about STEM PCK are addressed in Chaps. 3, 4, and 5 of this book.)

Teachers struggle to enact effective learner-centered approaches of integrated STEM education even after demonstrating a conceptual understanding of STEM instruction (Han, Yalvac, Capraro, & Capraro, 2015). Therefore, we adopted a model of Synthesize Qualitative Data for preparing preservice teachers for technology use. Tondeur et al. (2012) outlined a promising model to enhance teachers' STEM PCK. Their model includes six key connected themes: role models, reflection, (re)designing instruction, collaboration, authentic experiences, and feedback. First, a teacher educator must serve as a role model to show examples of STEM curriculum and instructional designs and then guide preservice and in-service teachers to observe, discuss, and reflect upon the integration of STEM curriculum and successful teaching strategies. After reviews of these examples, teachers need to (re)design STEM curriculum through collaboration. Then, teachers need to enact their STEM curriculum in authentic settings in order to develop their STEM PCK through ongoing and process-oriented feedback from student performance and peer suggestions.

1.6   Conclusion

This review of science education research on integrated STEM experiences identified promising approaches for promoting learning in and across STEM subjects and enhancing students' interest and identity related to STEM. Additionally, it highlighted that more research is needed for enriching our understanding of various aspects of STEM education such as instructional design, teaching practices, outcome measures, and teacher education and PD. To address the issues identified in the review, we proposed an instructional design framework for developing integrated STEM curricula and a PD model for preparing and elevating teachers' professional attributes and growth. The 5-step design framework stresses that it is important to identify and decide the core competencies and to generate activities aligned with the development of these core competencies. The teacher preparation model draws attention to the cultivation of teachers' STEM PCK through presenting role models and creating a learning environment to engage preservice teachers in an iterative process of designing, teaching, and reflecting. By which, they can appreciate the value of integrated STEM and gradually develop and consolidate their beliefs about and identities toward STEM education.

Acknowledgments This work was financially supported by the Institute for Research Excellence in Learning Sciences of National Taiwan Normal University from the Featured Areas Research Center Program within the framework of the Higher Education Sprout Project and Ministry of Science and Technology 107-2511-H-003-043-MY3 Project by the Ministry of Education in Taiwan.

References

References marked with an asterisk indicate the 26 studies included in the review

*Adedokun, O. A., Bessenbacher, A. B., Parker, L. C., Kirkham, L. L., & Burgess, W. D. (2013). Research skills and STEM undergraduate research students' aspirations for research careers: Mediating effects of research self-efficacy. *Journal of Research in Science Teaching, 50*(8), 940–951.

Allen, M., Webb, A. W., & Matthews, C. E. (2016). Adaptive teaching in STEM: Characteristics for effectiveness. *Theory into Practice, 55*(3), 217–224.

Breiner, J., Harkness, S., Johnson, C. C., & Koehler, C. (2012). What is STEM? A discussion about conceptions of STEM in education and partnerships. *School Science and Mathematics, 112*(1), 3–11.

*Brown, R. E., & Bogiages, C. A. (2019). Professional development through STEM integration: How early career math and science teachers respond to experiencing integrated STEM tasks. *International Journal of Science and Mathematics Education, 17*(1), 111–128.

Bryan, L. A., Moore, T. J., Johnson, C. C., & Roehrig, G. H. (2016). Integrated STEM educa-tion. In C. C. Johnson, E. E. Peters-Burton, & T. J. Moore (Eds.), *STEM road map: A frame-work for integrated STEM education* (pp. 23–37). New York, NY: Routledge/Taylor & Francis.

*Burgin, S. R., McConnell, W. J., & Flowers III, A. M. (2015). "I actually contributed to their research": The influence of an abbreviated summer apprenticeship program in science and engineering for diverse high-school learners. *International Journal of Science Education, 37*(3), 411–445.

*Carrier, S. J., Whitehead, A. N., Walkowiak, T. A., Luginbuhl, S. C., & Thomson, M. M. (2017). The development of elementary teacher identities as teachers of science. *International Journal of Science Education, 39*(13), 1733–1754.

*Chien, Y. H., & Chu, P. Y. (2017). The different learning outcomes of high school and college students on a 3D-printing STEAM engineering design curriculum. *International Journal of Science and Mathematics Education, 16*(6), 1047–1064.

*Dickerson, D. L., Eckhoff, A., Stewart, C. O., Chappell, S., & Hathcock, S. (2014). The examination of a pullout STEM program for urban upper elementary students. *Research in Science Education, 44*(3), 483–506.

Ejiwale, J. (2013). Barriers to successful implementation of STEM education. *Journal of Education and Learning, 7*(2), 63–74.

English, L. D. (2016). STEM education K-12: Perspectives on integration. *International Journal of STEM Education, 3*(1). https://doi.org/10.1186/s40594-016-0036-1.

English, L. D. (2017). Advancing elementary and middle school STEM education. *International Journal of Science and Mathematics Education, 15*(1), 5–24.

*Guzey, S. S., Ring-Whalen, E. A., Harwell, M., & Peralta, Y. (2019). Life STEM: A case study of life science learning through engineering design. *International Journal of Science and Mathematics Education, 17*(1), 23–42.

*Han, S., Capraro, R. M., & Capraro, M. M. (2015). How science, technology, engineering, and mathematics (STEM) project-based learning (PBL) affects high, middle and low achievers differently: The impact of student factors on achievement. *International Journal of Science and Mathematics Education, 13*(5), 1089–1113.

Han, S., Yalvac, B., Capraro, M. M., & Capraro, R. M. (2015). In-service teachers' implementation and understanding of STEM project based learning. *Eurasia Journal of Mathematics, Science and Technology Education, 11*(1), 63–76.

*Hughes, R. M., Nzekwe, B., & Molyneaux, K. J. (2013). The single sex debate for girls in science: A comparison between two informal science programs on middle school students' STEM identity formation. *Research in Science Education, 43*(5), 1979–2007.

*King, D., & English, L. D. (2016). Engineering design in the primary school: Applying stem concepts to build an optical instrument. *International Journal of Science Education, 38*(18), 2762–2794.

*Kitchen, J. A., Sonnert, G., & Sadler, P. M. (2018). The impact of college- and university-run high school summer programs on students' end of high school STEM career aspirations. *Science Education, 102*(3), 529–547.

*Korur, F., Efe, G., Erdogan, F., & Tunç, B. (2017). Effects of toy crane design-based learning on simple machines. *International Journal of Science and Mathematics Education, 15*(2), 251–271.

*Krogh, L. B., & Andersen, H. M. (2013). "Actually, I may be clever enough to do it": Using identity as a lens to investigate students' trajectories towards science and university. *Research in Science Education, 43*(2), 711–731.

*Lamb, R., Akmal, T., & Petrie, K. (2015). Development of a cognition-priming model describing learning in a STEM classroom. *Journal of Research in Science Teaching, 52*(3), 410–437.

*Means, B., Wang, H., Wei, X., Lynch, S. J., Peters, V. L., Young, V., et al. (2017). Expanding STEM opportunities through inclusive STEM-focused high schools. *Science Education, 101*(5), 681–715.

*Means, B., Wang, H., Young, V., Peters, V. L., & Lynch, S. J. (2016). STEM-focused high schools as a strategy for enhancing readiness for postsecondary STEM programs. *Journal of Research in Science Teaching, 53*(5), 709–736.

*Micari, M., & Light, G. (2009). Reliance to independence: Approaches to learning in peer-led undergraduate science, technology, engineering, and mathematics workshops. *International Journal of Science Education, 31*(13), 1713–1741.

*Micari, M., Van Winkle, Z., & Pazos, P. (2016). Among friends: The role of academic-preparedness diversity in individual performance within a small-group STEM learning environment. *International Journal of Science Education, 38*(12), 1904–1922.

National Academy of Engineering & National Research Council. (2014). *STEM integration in K-12 education: Status, prospects, and an agenda for research.* Washington, DC: National Academies Press. https://doi.org/10.17226/18612.

National Research Council. (2013). *Next generation science standards: For states, by states.* Washington, DC: National Academies Press. https://doi.org/10.17226/18290.

*Park, D. Y., Park, M. H., & Bates, A. B. (2018). Exploring young children's understanding about the concept of volume through engineering design in a STEM activity: A case study. *International Journal of Science and Mathematics Education, 16*(2), 275–294.

Penner, D. E., Giles, N. D., Lehrer, R., & Schauble, L. (1997). Building functional models: Designing an elbow. *Journal of Research in Science Teaching, 34*(2), 125–143.

*Prieto, E., & Dugar, N. (2017). An enquiry into the influence of mathematics on students' choice of STEM careers. *International Journal of Science and Mathematics Education, 15*(8), 1501–1520.

*Romine, W. L., & Sadler, T. D. (2016). Measuring changes in interest in science and technology at the college level in response to two instructional interventions. *Research in Science Education, 46*(3), 309–327.

Ruiz-Primo, M. A., Shavelson, R. J., Hamilton, L., & Klein, S. (2002). On the evaluation of systemic science education reform: Searching for instructional sensitivity. *Journal of Research in Science Teaching, 39*(5), 369–393.

*Sahin, A., Gulacar, O., & Stuessy, C. (2015). High school students' perceptions of the effects of international science olympiad on their STEM career aspirations and twenty-first century skill development. *Research in Science Education, 45*(6), 785–805.

Saxton, E., Burns, R., Holveck, S., Kelley, S., Prince, D., Rigelman, N., et al. (2014). A common measurement system for K-12 STEM education: Adopting an educational evaluation methodology that elevates theoretical foundations and systems thinking. *Studies in Educational Evaluation, 40,* 18–35.

*Schnittka, C. G., Evans, M. A., Won, S. G. L., & Drape, T. A. (2016). After-school spaces: Looking for learning in all the right places. *Research in Science Education, 46*(3), 389–412.

*Schütte, K., & Köller, O. (2015). "Discover, understand, implement, and transfer": Effectiveness of an intervention programme to motivate students for science. *International Journal of Science Education, 37*(14), 2306–2325.

Shaughnessy, J. M. (2013). Mathematics in a STEM context. *Mathematics Teaching in the Middle School, 18*(6), 324–327. https://doi.org/10.5951/mathteacmiddscho.18.6.0324.

Shernoff, D. J., Sinha, S., Bressler, D. M., & Ginsburg, L. (2017). Assessing teacher education and professional development needs for the implementation of integrated approaches to STEM education. *International Journal of STEM Education, 4*(1). https://doi.org/10.1186/s40594-017-0068-1.

*Tippett, C. D., & Milford, T. M. (2017). Findings from a pre-kindergarten classroom: Making the case for STEM in early childhood education. *International Journal of Science and Mathematics Education, 15*(Suppl 1), 67–86. https://doi.org/10.1007/s10763-017-9812-8.

*Todd, B., & Zvoch, K. (2017). Exploring girls' science affinities through an informal science education program. *Research in Science Education*. Advance online publication. https://doi.org/10.1007/s11165-017-9670-y.

Tondeur, J., van Braak, J., Sang, G., Voogt, J., Fisser, P., & Ottenbreit-Leftwich, A. (2012). Pre-paring pre-service teachers to integrate technology in education: A synthesis of qualitative evidence. *Computers & Education, 59*(1), 134–144.

Vasquez, J. A., Sneider, C., & Comer, M. (2013). *STEM lesson essentials, grades 3–8: Integrating science, technology, engineering, and mathematics*. Portsmouth, NH: Heinemann.

Chapter 2
The Potential of Arts-Integrated STEM Approaches to Promote Students' Science Knowledge Construction and a Positive Perception of Science Learning

Hye-Eun Chu, Yeon-A. Son, Hyoung-Kyu Koo, Sonya N. Martin and David F. Treagust

2.1 Introduction

This chapter presents an account of how the integration of the arts in the teaching of science, technology, engineering, and mathematics education (STEM) was conceptualized and implemented in a research project involving teachers and students in Australia and South Korea. It sets out the project's social constructivist framework and inquiry-based pedagogy, before giving some details of how arts- and culture-related content was integrated into teaching/learning activities in primary and secondary science classrooms. The project's outcomes provide grounds to not only argue for the beneficial effects of the science, technology, engineering, arts, and mathematics (STEAM) approach on student learning and perception of science but also indicate that there are some challenges.

The traditional approach to teaching science in schools has focused more on the transmission and accurate recall of scientific facts and theories than on student understanding of scientific concepts and their application in the real world. Science was—and still is—taught in isolation from other subjects, including mathematics; the emphasis in each subject was on knowledge acquisition and problem-solving (Song, 2004). Documented declines in student interest for studying science and increasingly negative student attitudes toward science are potential outcomes of this emphasis on

H.-E. Chu (✉)
Department of Educational Studies, Macquarie University, Sydney, Australia
e-mail: hye-eun.chu@mq.edu.au

Y.-A. Son
Department of Science Education, Dankook University, Yongin, Republic of Korea

H.-K. Koo · S. N. Martin
Department of Earth Science Education, Seoul National University, Seoul, Republic of Korea

D. F. Treagust
School of Education, Curtin University, Perth, Australia

© Springer Nature Singapore Pte Ltd. 2019
Y.-S. Hsu and Y.-F. Yeh (eds.), *Asia-Pacific STEM Teaching Practices*,
https://doi.org/10.1007/978-981-15-0768-7_2

17

the acquisition of isolated science and mathematics knowledge. The 1999 TIMMS Index of Students' Positive Attitude toward the Sciences indicates that only 10% of South Korean students and 27% of Australian students had a high level of positive attitude to science, compared to the international average of 40% (Martin et al., 2000). A similar picture emerges in the 2011 TIMSS survey of students' perceptions of science study: where only 11% of South Korean students and 25% of Australian students reported they enjoyed learning science (Martin, Mullis, Foy, & Stanco, 2012).

The aversion to science subjects has also been noted in national reports in South Korea where the number of applications to study STEM subjects at university in 2002 dropped to 27% from 43% in 1997 (Cho, Lee, & Park, 2003), indicating that the emphasis on knowledge transmission in traditional science classrooms has led to students perceiving science as difficult and uninteresting (Cho et al., 2003; Hong, 2016; Jon & Chung, 2013). In Australia, there is similar concern over the low number of students studying engineering, manufacturing, and STEM-related subjects at university, with about 8% of graduates in these fields compared to an OECD average of 14% (Singhai, 2017). The low interest among students in science and other STEM subjects is a particularly urgent problem when one considers the role of STEM-qualified, creative thinking scientists in maintaining a nation's technological competitiveness.

The trend in industry in developed countries today is a move from manufacturing to technology requiring human resources equipped with a combination of STEM knowledge, creativity, and critical thinking. These capabilities are needed to work toward building the industries of the twenty-first century that involve innovations such as artificial intelligence capability and 3D printing. In Australia, awareness of the need to equip citizens with new skills for new industries led to the government aiming to increase the "proportion of Australia's STEM experts that work in industry, business and the public sector outside of universities" (Department of Industry, Innovation & Science, 2015, p. 9). To achieve this aim, more students must be encouraged to study STEM subjects and build careers in STEM-related work. However, the Chief Scientist of Australia expressed concern at the decline in Australian schools "in the rates of participation in science subjects to the lowest level in 20 years" (Office of Chief Scientist, 2014, p. 11). There have been calls from previous chief scientists for STEM educators to "better engage students on STEM-related career pathways" (Taylor, 2016, p. 89). At the university level, the number of domestic undergraduates enrolled in science subjects in 2014 accounted for approximately 8% of total enrolments compared to 22% enrolled in arts subjects (Norton & Cakitaki, 2016).

Since the 1970s, South Korea's industry leaders have been aware that the nation needs scientists and engineers who are not only highly qualified in STEM areas but also capable of creative thinking and innovation because the country's manufacturing, shipbuilding, electronics, and semiconductor industries are competing with other technologically advanced countries in the world market (Cho et al., 2003; Lee, Jang, & Han, 1999). It became apparent that the school curriculum for science and other STEM subjects needed to be changed to respond to the call for a new-generation workforce knowledgeable in STEM and also capable of creative and critical thinking.

At the beginning of the twenty-first century, South Korea's Ministry of Education also became aware of the shortcomings in STEM education, particularly of the disconnect between the content knowledge taught in science and mathematics classes and students' interests as related to their real-life contexts (Ministry of Education, Science, and Technology [MEST], 2010). Another shortcoming noted was the limited opportunity in the school STEM curriculum for students to engage in creative design and problem-solving (MEST, 2010). Consequently, a changed approach to STEM education, particularly science education, was necessary.

Responding to the need to invigorate school STEM lessons with relevance to everyday life and provide scope for developing creative thinking, in 2009 MEST policy-makers and science educators revised the National Science Curriculum to integrate elements of the arts, technology, and/or engineering with the teaching/learning of science (MEST, 2009). The belief was that such integration would promote students' better understanding of science concepts and their appreciation of the connections between science knowledge and other school subjects. This concept of integrating the teaching/learning of science with other subjects came to be known as the STEAM approach. STEAM was a science education term originally used by Yakman, a teacher educator, in her thesis (2006). Yakman was the first to propose a framework for integrating the arts—visual arts, photography, performance, literature, and history—with STEM lessons (2008). The South Korean STEAM approach, which is implemented "without diluting content" (Marginson, Tytler, Freeman, & Roberts, 2013, p. 16), is informed by a theoretical framework comprising three principles: situating the teaching/learning of science in real-life problems/events, encouraging students to exercise creativity in devising their own solutions to a problem, and generating enthusiasm and positive attitudes toward doing science among students (Korean Foundation for the Advancement of Science & Creativity [KOFAC], 2012).

The shared goal of making science study attractive and relevant to students brought Australian and South Korean science educators together to collaborate on the creation of a STEAM program that would engage students and produce positive attitudes toward science.

2.2   The Australian–Korean STEAM Project

With a grant secured from the Australia–Korea Foundation, the researchers developed a STEAM program grounded in a social constructivist theory of learning and inquiry-based pedagogy. Applying a social constructivist perspective of learning, science concepts are presented in situations in which the concept plays a role in solving a problem or explaining a phenomenon. The theory holds that construction of knowledge by the students themselves is assisted by interaction with their peers and the teacher (Vygotsky, 1978, 1986); the teacher does not deliver knowledge to them. In a typical classroom situation, the social interaction that influences students' construction of knowledge takes the form of classroom and small-group discussions or synchronous and asynchronous online forums. Construction of knowledge is manifested in explanatory models (e.g., diagrams, written explanations) generated by students to articulate their pre- and post-instruction understanding of the target science concepts.

2.2.1 An Inquiry-Based Learning Process

Knowledge construction in the STEAM framework is guided by an inquiry-based teaching/learning process, which is rooted in social constructivism (Walker & Shore, 2015). In our program, the students begin with asking questions triggered by an experience in their social-cultural life—after which the teacher presents (e.g., a video clip of a light show) and students attempt to generate an explanation or solution to answer the questions. Students then engage in a hands-on or information and communication technology-mediated teaching/learning activity designed by the teacher to address incomplete or alternative conceptions manifested in the students' initial expositions. Research has indicated that a hands-on approach is more desirable than a teacher-talk methodology (Saunders, 1992). The experience gained from the hands-on teaching/learning activity enables students to collaboratively evaluate and revise their previous concepts, which were derived from prior unexamined experiences and may be incomplete or inaccurate (Saunders-Stewart, Gyles, Shore, & Bracewell, 2015), thus furthering their knowledge construction to arrive at a scientifically acceptable concept.

An arts-integrated STEM curriculum situates the teaching and learning of science concepts in a social-cultural context familiar to students, with the ultimate goal of showing them the relevance of science to their everyday life. An arts-related event or a social-cultural issue in a community's conversation, such as climate change (Jeong & Kim, 2015), can provide the context for inquiry into and exploration of the science that accounts for the experiential features of the event or issue. For example, the festival known as Vivid Sydney features a light show projected on the Sydney Opera House, which is an ideal cultural event for Australian students to inquire into and thus learn about light refraction and the mechanism by which eyes perceive images. Arts-related activities can also be a means for students to apply and deepen their understanding of a science concept. In the case of light refraction and the perception of light in the act of seeing, the arts-related activity may be the design and creation of quasi-holograms that engage students in applying the concepts of the angle of light refraction and the best location of the eye in relation to the refracted light.

Integration of the arts into science lessons delivered through an inquiry-based methodology provides opportunities for students to engage in problem-solving and the exercise of creative thinking (Aulls & Shore, 2008; Marginson et al., 2013). In producing an artifact (e.g., a quasi-hologram), students have to solve problems such as appropriately positioning the image to be projected in the hologram and choosing colors that will enhance the visibility of the image. The problem-solving involved in the production of an arts-related artifact requires students to construct knowledge as they attempt to apply the target science concepts (e.g., light refraction). At the same

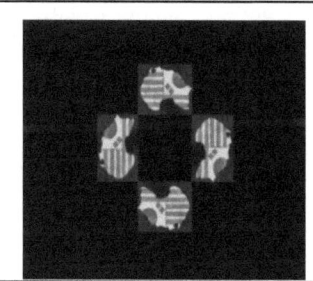 Creativity: South Korean and USA flags superimposed on the shape of the Australian continent in a quasi-hologram.

Fig. 2.1 Creative thinking in a STEAM learning activity

time, creative thinking is called into play because art generally communicates the creator's personal message or projects the creator's uniquely personal perspective. For example, one student designed a hologram that superimposed parts of the South Korean and USA flags on the shape of the Australian continent as a symbol of her multicultural background: born in South Korea, she had studied in the US and was now living in Australia (Fig. 2.1).

2.2.2 Group Knowledge Construction via Successive Explanatory Models

Knowledge construction through interaction with the teacher and with fellow students is an ongoing process during a STEAM lesson and even after the lesson through asynchronous interaction on an online platform where students post their artwork. The knowledge construction process in science learning consists of generating, evaluating, and modifying mental models that target the science concept (Clement & Rea-Ramirez, 2008; Windschitl, 2004). A student's articulation of his/her initial mental model of a science concept is facilitated through social interaction by listening to classmates' perceptions and exploring prior knowledge related to the concept. This process encourages the student to express his/her prior knowledge and experience. The initial model is generally the student's explanation of a scientific phenomenon or an answer to a question about the phenomenon (e.g., Why are there different seasons?). Students then engage in a group hands-on activity that the teacher has designed to address inaccuracies in students' initial models. Observations and new learning from the hands-on activity usually create some awareness in students of the inadequacy of their initial model, which then leads to its revision. The ensuing group discussion allows students to present their revised models and then negotiate a group model by either selecting one group member's creation or jointly constructing a revised model with what they consider desirable features from the models of other group members. Guided by the teacher and using equipment or knowledge resources available to them, each group evaluates its model for how well it explains a scientific phenomenon or answers satisfactorily the question posed at the beginning of the lesson. The group evaluation and modification of the model are effected through students questioning each other, sharing ideas, and using evidence from the hands-on activity to support their views. Throughout the process of teaching/learning, students are engaged in joint knowledge construction through social interaction.

The process of joint evaluation and modification of students' explanatory models leads to growth of understanding of the science concept (Clement, 2000; Koo, Chu, Martin, & Choe, 2017) and movement toward an acceptable scientific model. As evidence of such conceptual growth, this study presents three versions of a model (Fig. 2.2) explaining why different parts of the world (Australia & South Korea) have opposite seasons (winter & summer) at the same time of the year. Figure 2.2a, produced by an individual student after group sharing of ideas on seasons in different countries, accounts for the different seasons by attributing it to the Earth's rotation on its axis. This student then engaged in a group hands-on activity in which students shining a torch over a globe to represent the Sun saw the effect of the tilting of the

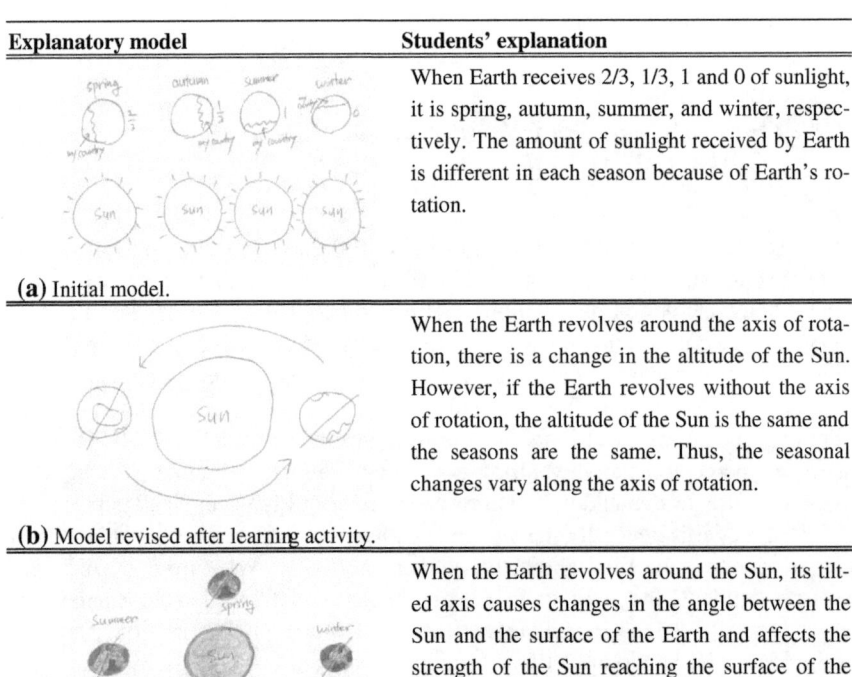

Explanatory model	Students' explanation
(a) Initial model.	When Earth receives 2/3, 1/3, 1 and 0 of sunlight, it is spring, autumn, summer, and winter, respectively. The amount of sunlight received by Earth is different in each season because of Earth's rotation.
(b) Model revised after learning activity.	When the Earth revolves around the axis of rotation, there is a change in the altitude of the Sun. However, if the Earth revolves without the axis of rotation, the altitude of the Sun is the same and the seasons are the same. Thus, the seasonal changes vary along the axis of rotation.
(c) Final model after testing of previous model and group discussion.	When the Earth revolves around the Sun, its tilted axis causes changes in the angle between the Sun and the surface of the Earth and affects the strength of the Sun reaching the surface of the Earth. Also, the seasons of South Korea and Australia, which are located in the northern and southern hemisphere respectively, are opposite at the same time of year.

Fig. 2.2 Development of conceptual understanding in successive explanatory models

Earth's axis on the Sun's altitude in two different parts of the Earth (Australia & South Korea). The ensuing group discussion led the student and group members to produce the model in Fig. 2.2b. The model now attributed the occurrence of different seasons to the Earth's revolution, the Earth's tilted axis (the students' drawing shows revolution around the Sun and the Earth's axis tilted), and changes in the altitude of the Sun. Although the drawing represents the concept correctly, the students' explanatory notes indicate an incomplete grasp of the concept. Assisted by prompts from the teacher, the group next tested the validity of the model in Fig. 2.2b by first measuring the length of the shadow of a short stick placed on the globe's northern hemisphere and then on the southern hemisphere; the group worked out that the temperature in a location in each hemisphere is influenced by the altitude of the Sun. After this test and group discussion of the results, the group modified Fig. 2.2b to produce their final model in Fig. 2.2c. The improvement in Fig. 2.2c lies in the accuracy of the explanation of seasonal change. Figure 2.2b strings together words representing bits of the whole concept (e.g., axis of rotation, altitude of the Sun); in Fig. 2.2c, the concept of why there are different seasons in different parts of the world at the same time of year is accurately explained: *When the Earth revolves around the Sun, its tilted axis causes changes in the angle between the sunlight and the surface of the Earth.* Through social interaction and joint construction of knowledge, these science learners developed increasingly accurate mental models of science concepts.

2.2.3 STEAM as the Teacher's Aid

Aside from enabling students to jointly construct models of science concepts through social interaction, the STEAM approach provides the means by which teachers can identify alternative conceptions and track students' progress in developing scientifically acceptable concepts so that subsequent lessons can be planned to address gaps in students' understanding. Teachers' analysis of students' explanatory models, written explanations, and learning artifacts can reveal missing elements in a student's mental model or a fragmented understanding in which relations between the components of a concept have not been correctly worked out.

Figure 2.3 shows an example of students' developing understanding of the concept of the role of the Earth's revolution in the occurrence of different seasons in the northern and southern hemispheres at the same time of year. The alternative concept in Fig. 2.3a is that the Earth's rotation on its axis is a contributory cause of the seasons in different places (South Korea & Australia). This drawing of the Earth's path around the Sun shows the Earth's axis tilting in different directions, indicating an absence of the concept that seasons are caused by the Earth's axis tilting in an unchanging direction during the Earth's revolution around the Sun. Following the production of individual initial models like Fig. 2.3a, the students viewed a video of the Earth's revolution around the Sun, listened to the teacher's explanation of the difference between the Earth's rotation and its revolution, participated in an activity in which the strength of the Sun's rays on different parts of the Earth was simulated

Explanatory model	Students' explanation
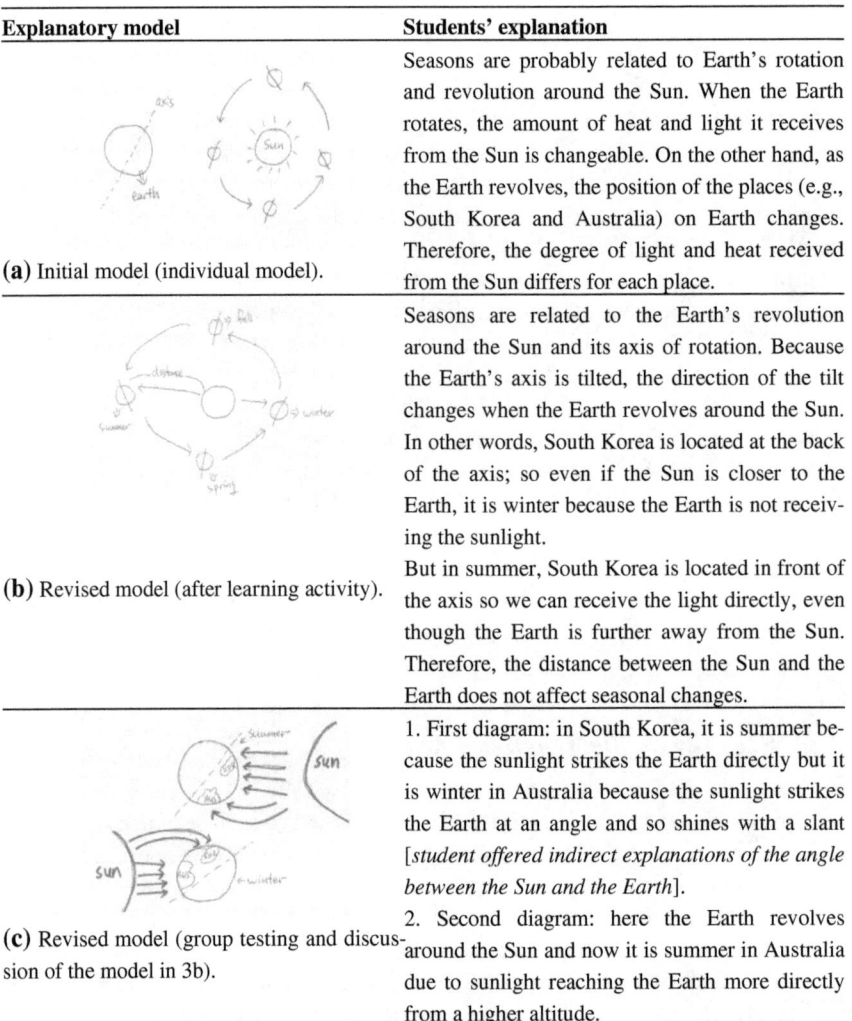 **(a)** Initial model (individual model).	Seasons are probably related to Earth's rotation and revolution around the Sun. When the Earth rotates, the amount of heat and light it receives from the Sun is changeable. On the other hand, as the Earth revolves, the position of the places (e.g., South Korea and Australia) on Earth changes. Therefore, the degree of light and heat received from the Sun differs for each place.
(b) Revised model (after learning activity).	Seasons are related to the Earth's revolution around the Sun and its axis of rotation. Because the Earth's axis is tilted, the direction of the tilt changes when the Earth revolves around the Sun. In other words, South Korea is located at the back of the axis; so even if the Sun is closer to the Earth, it is winter because the Earth is not receiving the sunlight. But in summer, South Korea is located in front of the axis so we can receive the light directly, even though the Earth is further away from the Sun. Therefore, the distance between the Sun and the Earth does not affect seasonal changes.
(c) Revised model (group testing and discussion of the model in 3b).	1. First diagram: in South Korea, it is summer because the sunlight strikes the Earth directly but it is winter in Australia because the sunlight strikes the Earth at an angle and so shines with a slant [*student offered indirect explanations of the angle between the Sun and the Earth*]. 2. Second diagram: here the Earth revolves around the Sun and now it is summer in Australia due to sunlight reaching the Earth more directly from a higher altitude.

Fig. 2.3 Tracking students' concept development

with a torch, and engaged in a class discussion of their new learning. The student who created the model in Fig. 2.3a produced Fig. 2.3b, thus demonstrating that he no longer believed in the Earth's rotation as one of the causes of seasonal change. The student's diagram in Fig. 2.3b reflects a revised mental model in which the Earth's revolution around the Sun is the reason for the occurrence of winter or summer in the two hemispheres at different times of the year. However, the model in Fig. 2.3b retains the alternative conception: the belief that the tilt of the Earth's axis changes direction as the Earth revolves around the Sun—an erroneous concept that invalidates the model that the Earth's tilted axis causes varying amounts of sunlight to be received at different locations on the planet as it revolves.

The inconsistent direction of tilt is rectified in Fig. 2.3c, indicating that in their group discussion the students have learnt that the tilt in the Earth's axis remains unchanged during the Earth's revolution and that this is what causes the angle between the Sun's rays and the Earth to be smaller or larger, depending on whether a country is located in the northern or southern hemisphere. But the curved lines representing the Sun's rays in Fig. 2.3c may indicate a concept still to be acquired: that of sunlight reaching the Earth as rays that are almost completely parallel because the Sun is 109 times larger than the Earth and nearly 150 million kilometers away. The incremental gains in the students' explanatory models in Fig. 2.3 illustrate how the STEAM approach allows teachers to identify students' alternative or incomplete concepts and provides teachers with cues to the planning of subsequent lessons designed to address the gaps in students' conceptual understanding.

2.2.4 STEAM as an Approach for Promoting Students' Science Learning and Interests

The STEAM approach developed in this study applied an inquiry-based teaching methodology, a mode of instruction that research studies have confirmed as conducive to better science learning (Chang & Mao, 2010; Furtak, Seidel, Iverson, & Briggs, 2012). One key practice of inquiry-based teaching is to start from questions generated by students, which they then investigate to find the answers. In integrating arts-related events or products into science lessons, the STEAM approach provides a context that awakens interest and curiosity in students, naturally leading them to inquiry questions (e.g., How do they make the colors in Vivid Sydney appear so bright on the Sydney Opera House?). Since the students themselves generate the questions, they feel a sense of ownership and are motivated to work at investigation and joint knowledge construction (Bevins & Price, 2016) as they create and critique successive explanatory models. In one of the STEAM lessons, after the students had viewed video clips of Vivid Sydney and a South Korean light festival at Gyeongbokgun Palace, they observed that the projected images created by light in Vivid Sydney were brighter than those at Gyeongbokgun Palace. Further discussion, guided by the teacher, led to the two student-generated inquiry questions below with the related science concept in parentheses.

- What is the connection between the brightness of light and the type and nature of the materials used to construct the building at Sydney Opera House/South Korea's Gyeongbokgun Palace? (Refraction of light depends on smoothness, reflectiveness, and color.)
- Why do we have light festivals only at night? (The human eye's perception of a light source is affected by ambient light.)

These inquiry questions directed the teacher's planning of investigative activities that guided students to construct the science concepts shown in parentheses and answer the questions. The marrying of the STEAM approach and inquiry-based teaching methodology makes science teaching/learning more relevant to students and thus more effective than a pedagogy that treats subjects as "silo disciplines" (Yakman & Lee, 2012, p. 4).

In addition to engaging students in problem-solving and joint knowledge construction with their peers and enabling teachers to monitor students' conceptual development, the STEAM approach appears to have had a positive effect on students' perceptions about science and science study. The first type of attitudinal change found in our study was the perception of the relevance of science to everyday social-cultural life, including arts-related experiences like light festivals. One Australian student (L), interviewed after a program of STEAM lessons, said, "I actually never thought that this stuff, for example, light festivals, holograms, have anything to do with science. It's impressive how science can be used for art." What Student L has learnt from the STEAM activities is that science is not merely content knowledge confined to textbooks and science laboratories but plays a role in the creation and enjoyment of the arts. More than 90% (32/35) of interviewed students expressed similar sentiments, indicating a departure from a segregated perception of STEM courses (Brown, Brown, Reardon, & Merrill, 2011) and a growing realization of the connection between scientific concepts and the students' life experiences outside the science classroom and laboratory.

The second type of attitudinal change observed in the students was from perceiving science as a difficult subject to perceiving it as engaging. Student L admitted to tuning out in previous science lessons because the content was "very complicated, [but] this STEAM program is exciting [because] you realize how light works" (referring to the six lessons on light). Student K said, "I thought it [light] was just a unit to learn and pass, but after this lesson I became very interested in light." Furthermore, Student K's pre-STEAM perception of science was that it was a subject in which students are taught what to know (Morrison, 2006) and what to correctly reproduce to pass a course. After participating in collaborative attempts to apply a group-created explanatory model of how human eyes perceive objects to the production of an artistic artifact (a quasi-hologram), Student K began to see science as intrinsically interesting and engaging. More than 94% (33/35) of interviewed students reported that the STEAM lessons impacted similarly on their perception of science and/or science learning.

2.3   Concerns and Challenges

Notwithstanding the benefits presented above, concerns have surfaced about the feasibility of implementing STEAM in school as part of a regular science curriculum. In South Korea, where STEAM has been implemented nationwide in primary and secondary schools since 2011 (MEST, 2011a; MEST, 2011b), teachers have reported challenges relating to integrating the arts with science and mathematics lessons. The main challenges that have surfaced or are evident from the literature are as follows:

1. The difficulty STEAM teachers face in integrating arts-related content with the teaching of core science concepts (Choi, Lee, & Noh, 2015; Sim, Lee, & Kim, 2015; Son, Jung, Kwon, Kim, & Kim, 2012);
2. An incomplete understanding among teachers of the underlying principles and goals of STEAM (Park, Byun, Sim, Baek, & Jeong, 2016);
3. Insufficient curriculum time for the teaching of core science concepts alongside integration of the arts into the lessons (KOFAC, 2012); and
4. The mismatch in the national examinations between the creative problem-solving encouraged i n STEAM lessons (Noh & Paik, 2014) and the examination's emphasis on accurate recall of science knowledge (Ministry of Education, 2015).

Integrating arts-related content with the teaching of core science concepts poses a number of challenges for STEM teachers applying the STEAM approach for the first time. The foremost challenge is the selection of arts- or culture-related content (artifact or event) that could be used to help students develop an understanding of the core science concept (Chu, Martin, & Park, 2018; Lim, 2012). Next, having selected the arts or culture content, there is the question of how to use the arts/culture event in teaching/learning activities so that students appreciate the role and relevance of the science concept in a familiar social-cultural experience (Chu et al., 2018; KOFAC, 2018). The lack of experience and skill among teachers when it comes to incorporating arts activities that can develop students' understanding of a science concept has led to teachers including arts-related activities simply for their own sake, without them playing any part in demonstrating the target science concept (Park et al., 2012; Park et al., 2016). The problem is particularly acute for secondary school science teachers, the majority of whom majored in science at university and may feel inadequate in their knowledge of non-science fields such as literature, visual arts, and history. Collaboration between science and non-science teachers, which could solve the problem, has been very low according to a survey of 300 South Korean STEAM teachers (KOFAC, 2012), probably due to time and structural constraints.

One factor contributing to teachers' uncertainty about how to meaningfully integrate the arts with science lessons is that they do not fully understand the underlying principles and goals of STEAM. What may be missing in many teachers' conceptualization of STEAM is its goal to enable students to answer cross-cutting questions (Lee, 2012; Lee, Kim, & Lee, 2013) that are answered by applying knowledge and skills from the arts, STEM, and other non-science disciplines. The underlying principle of STEAM is the use of experiences and knowledge from non-STEM subjects (i.e., visual arts, literature, history, social studies) as a medium and/or a context for developing understanding of science concepts (Chu et al., 2018; Korea Educational & Development Institute, 2012; Sim et al., 2015). Ideally, the integration of science and non-science disciplines should be understood as focused on helping students to construct new knowledge without being restricted by the boundaries between disciplines, such as those separating science and the visual arts (Park & Shin, 2015; Rennie, Venville, & Wallace, 2012). Some teachers understand STEAM as the inclusion of an arts/culture-related activity purely for the purpose of engaging student interest at the beginning of a lesson (Park, Chu, & Martin, 2017). But this approach does not achieve the goal of STEAM because it does not engage students in applying knowledge from science and non-science disciplines, including the arts, to solve or understand real-world problems (e.g., how to solve the global epidemic of overweight and obesity, see https://www.who.int/nutrition/topics/obesity/en/, that involves the interplay of biochemistry, social studies, and psychology).

The third challenge in implementing STEAM successfully is the perception that the time consumed by the teaching of core science concepts in the typical science curriculum leaves little time left for arts/culture-related content and learning activities (Lee & Shin, 2014; Lim, Kim, & Lee, 2014; Shin & Han, 2011). Science teachers believe that there is already a heavy load of science content to teach without including the arts in science lessons (Park et al., 2017; Yoo, Hwang, & Han, 2016). This concern suggests a dichotomous view of science and the arts, pointing perhaps to an incomplete understanding of the principle of STEAM. There is also the concern that integrating the arts with science lessons spells an additional demand on the teacher's lesson preparation time (Im, Kim, & Lee, 2014; Park et al., 2014) since teachers would need to source relevant content from other disciplines and plan teaching/learning activities that involve students in applying science concepts when explaining an arts/culture-related event or in creating an artistic artifact.

The fourth constraint in the implementation of STEAM is the mismatch between the creative arts/culture-situated problem-solving in STEAM lessons and the emphasis on accurate recall of science knowledge in the national examinations in countries like South Korea. In STEAM lessons, classroom activities engage students in the creative design of possible solutions they generate in response to a real-world problem as well as the testing of solutions to find the most suitable solution (Baek et al., 2011). But science examination questions in national examinations (e.g., South Korea's College Scholastic Ability Test) tend to require students to demonstrate knowledge of scientific terms and facts by selecting the correct response in multiple-choice questions. Teachers could, therefore, be conflicted about whether to teach for the examination or to engage students in creative problem-solving through the STEAM approach. Teachers who have reason to be anxious about their students' performance in the national examinations might be reluctant to spend class time on STEAM activities, even though, in the case of South Korea, the Ministry of Education has mandated that 20% of curriculum time be allocated to STEAM (KOFAC, 2012).

2.3.1   Addressing the Challenges

Although the challenges presented above are understandable, given the social and educational structures that have supported the traditional teacher-centered, knowledge-focused approach to teaching science, there are ways to overcome the barriers to wider acceptance of STEAM and its successful implementation. The educational literature and the practices from this STEAM project serve as a basis to propose some ways of overcoming the difficulties implicit in the widespread implementation of STEAM.

The difficulty teachers face regarding how they should integrate the arts and which aspects of the arts into science lessons is rooted in the multifaceted references encased in the term *art/arts* from a restricted definition encompassing only the visual arts (i.e., painting, drawing, photography, sculpture) to a more inclusive definition covering the performing arts (i.e., dance, music, theater) and the liberal arts and humanities (Herro & Quigley, 2017; Quigley, Herro, & Jamil, 2017). Teachers who have not studied art/arts in the above senses are quite lost as to what form arts integration could take. Even non-arts graduate teachers who recognize the benefits of arts-based learning may struggle to find effective strategies for integrating the arts with other subjects (Rabkin & Hedberg, 2011). Furthermore, many teachers think of arts integration as the inclusion of an arts product in the science lesson rather than engaging students in the process of artistic inquiry (Sullivan, 2006), which includes exploration, risk-taking, flexible thinking, and learning from mistakes (Hetland, 2013; Hetland, Winner, Veenema, & Sheridan, 2007; Sawyer, 2006; Sawyer & DeZutter, 2009). Integrating the arts with science teaching/learning requires an understanding of the role of an artistic inquiry process in building conceptualizations of science and appreciation of the relevance of science in social-cultural life.

What could help teachers to select arts/culture-related content for integration in science lessons would be professional development (PD) programs in which teachers collaborate in exploring and selecting arts content (events/artifacts) that can be used to demonstrate the working or application of particular science concepts. A problem-solving approach might be adopted with posing questions such as: What arts/culture event(s) would involve students in explaining and/or applying the concept of light refraction and the human perception of light? The next step would be to discuss, and perhaps practice, strategies to steer student discussion of an arts/cultural event to the asking of questions that would lead to an inquiry into the science underlying a phenomenon students have observed in the arts/cultural event. The STEAM project reported here involved teacher-guided students' comments on video clips of light shows toward asking questions like: Where is the light coming from? The PD program should focus on teachers discussing strategies to encourage students to propose models to answer their inquiry question and account for the phenomenon they have just viewed, and then to test and revise the models.

The following example illustrates such a PD activity. Suppose students are shown visual images (e.g., photographs, paintings, video clips) of Christmas cards in England (with the traditionally attired Santa) and in Australia (Santa in shorts and thongs). The teachers would discuss and practice strategies to direct students' attention to the differences in the two images, which should then trigger the question to start the scientific inquiry (i.e., Why is Christmas happening in different seasons?). The inquiry questions arising from students' observations of phenomena would lead to conjectures based on students' existing knowledge when the teacher asks them to attempt an explanation of the phenomenon they have observed (i.e., Christmas occurring in two different types of climatic conditions in two countries on the same day). The PD session would show teachers how to plan hands-on teaching/learning activities that

address the deficits, as indicated in students' initial models, in students' conceptualization of the core science concept that accounts for the phenomenon, and then how
to guide students to test the validity of their explanatory model and to revise it after
they have engaged in the hands-on activity.

This example encapsulates the process of doing science: observing, model construction, investigating (more observing), analyzing results, model testing/revising,
and drawing conclusions (Clement, 2000; Miller, 1989). The teachers would collaborate to design an arts-integrated teaching/learning activity to help students acquire
the concept of the Earth's revolution with reference to different seasons in the two
hemispheres. For example, they might show students in classrooms in the southern
hemisphere advice from a real estate website that north-facing houses are the most
desirable, a cultural practice that allows more sunlight into the house in winter when
the Sun is at its lowest. Teachers would then help students to work out the science
underlying this advice by having students measure the angle of the Sun's rays in
winter/summer to observe the low altitude of the rays in winter. The PD process of
collaborating in selecting arts content and planning ways of arts integration to fulfill
the purpose of teaching a core science concept would start teachers on developing
their own expertise in arts integration.

The challenge posed by teachers' partial understanding of the meaning and purpose of STEAM is addressed in the PD program proposed above. When teachers use
a cultural practice (e.g., the preference for north-facing homes) as a context for investigating the effect of the Earth's revolution and the Earth's tilted axis on the angle
of the Sun's rays at particular times of the year, they see the role of the arts/culture
component in students' construction of new knowledge (in our example, knowledge
of why different seasons occur in the northern/southern hemispheres at the same time
of year). In this STEAM project, teachers were shown how to use a quasi-hologram
designing activity (the arts component) to engage students in applying the science
concepts of light refraction and the human eyes' and brain's perception of images created by light refraction. This activity helped teachers to understand the role of the arts
(e.g., hologram design, choice of design/colors) in developing students' conceptual
understanding of science.

The PD programs might also be designed for teachers to experience the process
of exploring answers to a cross-cutting question (Lee, 2012) by investigating the
question from the perspective of a few disciplines, science, and non-science. For
instance, in the topic of food and digestion in the New South Wales science syllabus
for Years 7–8 (New South Wales Education Standards Authority, 2018), a cross-
cutting inquiry question can be: How can the problem of obesity, widespread in many
countries, be tackled? Teachers might explore answers to this question by looking
into the attractiveness of refined or junk foods (e.g., marketing strategies, creative
filmed advertising), the physiological effects of these foods on the human body
(e.g., biology, biochemistry), and evidence of the harm from abandoning traditional
diets for refined, packaged foods (e.g., sociology, anthropology). The process of
seeking answers to a real-world problem by applying interdisciplinary knowledge
and practices would demonstrate to teachers how arts- and culture-related knowledge
and experiences could be made an integral part of the teaching of science. They would

then see the potential of STEAM to enrich and expand the scope of STEM education (Korean Educational Development Institute, 2012; Taylor, 2016).

The time entailed in teachers' preparation of STEAM teaching/learning activities and in conducting the activities in the classroom has been described as a burden by teachers who have taught in a STEAM program (Cho, Kim, & Huh, 2012). Many studies of STEAM teachers' experiences in South Korea, where STEAM has been implemented in 20% of the science curriculum in most schools since 2011, mention insufficient time as an obstacle to its effective implementation (Lee & Shin, 2014; Lim et al., 2014; Shin & Han, 2011). The burden of lesson preparation time and effort might be alleviated to some extent through collaboration between science teachers and colleagues teaching other subjects, a solution that has been attempted in South Korean schools, but collaboration has proved to be difficult because differences in disciplinary culture and discourse interfere with communication (Han & Lee, 2012; Noh & Paik, 2014). In primary schools, where the general practice is for one teacher to teach all or most subjects, the integration of the arts and other non-science subjects in science lessons may be easier. The primary school teachers in this STEAM project were more open to the idea of arts integration compared to their secondary school counterparts, possibly because the education of primary school teachers involves exposure to pedagogy across subjects.

Beyond the primary school, it must be acknowledged that it is unrealistic to ask teachers to teach their own subject and, on top of that, require them to provide input into the teaching of a colleague's subject. One approach to addressing this problem is to restructure the school curriculum to remove or blur the boundaries between subjects and create a curriculum organized around real-world problems or issues. In an interdisciplinary, problem-centered curriculum, students would process and apply knowledge and methods from two or more disciplines to solve a problem (e.g., how to resolve widespread obesity), explain a phenomenon (e.g., why is Christmas in winter/summer in the northern/southern hemisphere?), or view a situation or issue from different perspectives so as to "raise a new question in ways that would have been unlikely through a single discipline" (International Baccalaureate Organization, 2005–2019, para. 2). From the perspective of the science teacher, instead of planning a science lesson and then seeking help from busy colleagues teaching other subjects on the content and method of integrating arts/culture-related activities into the lesson, the science teacher can plan and teach a lesson collaboratively with colleagues from different disciplines. The number of disciplines involved would depend on the nature of the problem/situation or phenomenon. Versions of the type of synthesized curriculum proposed here are practiced in some schools, for example, the Liverpool Girls' High School (2019) and International Baccalaureate (2005–2019) schools.

Another way of organizing an integrated curriculum, which has been implemented in Daeyeon High School in South Korea, is to select STEAM-relevant subjects (e.g., home economics, art, and music—subjects that can be integrated into the teaching of science, mathematics, and technology) and allocate a maximum of 20% of each subject's curriculum to STEAM (Im & Lee, 2012). For STEAM lessons, teachers select key topics from science, mathematics, or technology that hold concepts that are applicable in one or more non-STEM subject. For example, the science topic of neutralization reaction has relevance in home economics because it can explain why certain household substances (e.g., salt) are recommended for removing stains (e.g., wine) from fabric. Science and home economics teachers collaboratively plan the STEAM lesson to enable students to understand the science underlying the use of different cleaning agents in the topic of cleaning in home economics. Lesson delivery may involve the two subject teachers co-teaching one or more lessons, or teaching consecutive lessons, with the science teacher conducting learning activities to demonstrate neutralization reaction, followed by the home economics teacher engaging students in a problem-solving activity requiring students to select effective cleaning agents in a laundry scenario. This integrated curriculum can, to some extent, reduce the stress science teachers experience at having sole responsibility for selecting and integrating arts/culture-related content in the STEAM approach.

The mismatch between creative arts/culture-situated problem-solving in STEAM and the emphasis on accurate reproduction of science knowledge might be addressed with school-based assessment tasks such as projects that involve the application of core science concepts to find answers to an inquiry question or creating a solution for a problem. School-based assessment in the form of projects or small-scale investigations can situate inquiry questions or the problem to be solved in students' social-cultural experience and environment, and provide scope for students to engage in systems thinking (Arnold & Wade, 2015). Systems-thinking is a way of thinking that takes into account the interconnectivity between knowledge in different disciplines and develops an understanding of the complexity of situations and problems. In this STEAM project, formative assessment provided students with opportunities to create, review, and revise explanatory models to account for phenomena in their social-cultural life (e.g., celebrating Christmas in winter/summer). There is no reason to think that the same testing strategy cannot be used for summative assessment. But assessment that engages students in the process of posing questions; generating, testing, and revising tentative solutions or explanatory models; and exploring complexity requires work over a period of time and is not feasible in traditional 2- or 3-h written examinations. School-based assessment is already a feature in the assessment programs at South Korean schools implementing the STEAM approach in 20% of the curriculum (KOFAC, 2019). However, in countries like Singapore and South Korea where there is time-honored trust in national examination scores as the fairest method of selecting students for university courses, school-based assessments are regarded with a skepticism that is difficult to overcome (Shin, Ahn, & Kim, 2017). To raise stakeholders' trust in the validity of school-based assessment, a nationally recognized assessment audit of school-based assessment tasks and students' products can be established such as the School-based Assessment Audit used in the Australian state of Victoria. This School-based Assessment Audit aims to protect the integrity of the Victorian Certificate of Education, which comprises school-based assessment and state-level examinations. The annual audit checks the assessment information given in a school to its students, its marking schemes and criteria, and samples of student work to determine whether the principles and standards set by the Victorian Curriculum and Assessment Authority (n.d.) have been met. A similar system of audit could assure parents and university admission authorities that school-based assessment STEAM tasks have been set, administered, and evaluated according to the principles and standards of a national assessment board.

The concerns and challenges raised by teachers, teacher educators, education department/ministry officers, and education policy-makers are valid and deserve attention. But they are not insurmountable if there is commitment from teachers, teacher educators, and education policy-makers to understand the pedagogical purpose of STEAM and act to achieve that purpose—to help science learners appreciate the role and relevance of science and other STEM subjects in their social-cultural life. There is a need for time and STEAM-experienced personnel to be allocated to PD programs that are designed to equip teachers with an understanding of the principles of STEAM and the strategies for integrating arts-/culture-related events with science teaching. There is also a need for willingness among teachers and curriculum designers of all subjects to blur or remove the boundaries between subjects so that interdisciplinary inquiry and teaching can occur in schools. Finally, there is a need for measures to develop trust in school-based assessments such that the results from such assessments are regarded on a par with those of national examinations for the purpose of selection for university study.

2.4 Conclusion

This chapter has provided some evidence of the potential of the STEAM approach to help students to construct science concepts, acquire scientific ways of thinking, and build a positive perception of science and science study. It began with an account of the theoretical grounding of the Australian–South Korean STEAM project and the inquiry-based pedagogy applied in the STEAM lessons. The social constructivist theoretical framework and the inquiry-based teaching methodology were illustrated with examples of topics used for the project and of students' work. This material also evidenced the beneficial effects of the STEAM approach on the teaching/learning process and on students' perceptions of science and learning science.

As acknowledged in the preceding section, concerns and challenges that throw doubt on the feasibility of the widespread implementation of STEAM have emerged in countries where STEAM has been adopted as an alternative science curriculum. The challenges, such as the difficulty faced by science teachers in selecting and integrating arts- and culture-related content with science lessons, can be managed with effort and willingness on the part of teachers and curriculum designers to challenge and change traditional mindsets about the boundaries between science and non-science disciplines. The potential benefits of STEAM on student science learning would outweigh the intensive effort to bring about mindset and curricular changes. A firm belief in the potential of STEAM to improve outcomes in science education is an important reason the South Korean Ministry of Education has recommended that schools implement STEAM in the science curriculum.

Acknowledgements This work was supported by the Australian Government's Department of Foreign Affairs and Trade (Australia–Korea Foundation, AKF-2015 Grant 0098), by the Macquarie University New Staff Grant (GT-00058), and by the National Research Foundation of Korea Grant funded by the South Korean Government (NRF-2016S1A3A2925401).

References

Arnold, R. D., & Wade, J. P. (2015). A definition of systems thinking: A systems approach. *Procedia Computer Science, 44,* 669–678.

Aulls, M. W., & Shore, B. M. (2008). *Inquiry in education: The conceptual foundations for research as a curricular imperative* (Vol. 1). New York, NY: Routledge.

Baek, Y.-S., Park, H.-J., Kim, Y., Noh, S.-G., Park, J.-Y., Lee, J., et al. (2011). STEAM education in Korea. *Journal of Learner-Centered Curriculum and Instruction, 11*(4), 149–171.

Bevins, S., & Price, G. (2016). Reconceptualising inquiry in science education. *International Journal of Science Education, 38*(1), 17–29.

Brown, R., Brown, J., Reardon, K., & Merrill, C. (2011). Understanding STEM: Current perceptions. *Technology and Engineering Teacher, 70*(6), 5–9.

Chang, C., & Mao, S. (2010). Comparison of Taiwan science students' outcomes with inquiry-group versus traditional instruction. *Journal of Educational Research, 92*(6), 340–346.

Cho, H.-D., Lee, J.-W., & Park, J.-M. (2003). *A study on the realization of Korea's scientific and technological manpower in the 21st century: Key tasks and policy implications* [21 세기과학기술인력 강국실현: 핵심과제와 정책방안]. Seoul, KR: Science & Technology Policy Institute.

Cho, H.-S., Kim, H., & Huh, J.-Y. (2012). *Understanding converged human resources training (STEAM) through field application cases* [현장적용 사례를 통한 융합인재교육(STEAM)의 이해. Seoul, KR: Korean Educational Development Institute and Korea Foundation for the Advancement of Science & Creativity.

Choi, S., Lee, J., & Noh, T. (2015). A case study of preservice secondary science teachers' demonstration of STEAM lessons. *Journal of the Korean Association for Science Education, 35*(4), 665–676.

Chu, H. E., Martin, S., & Park, J. (2018). A theoretical framework for developing an intercultural STEAM program for Australian and Korean students to enhance science teaching and learning. *International Journal of Science and Mathematics Education.* Advance online publication. https://doi.org/10.1007/s10763-018-9922-y.

Clement, J. J. (2000). Model based learning as a key research area for science education. *International Journal of Science Education, 22*(9), 22–29.

Clement, J. J., & Rea-Ramirez, M. A. (2008). *Model-based learning and instruction in science.* Dordrecht, NL: Springer.

Department of Industry, Innovation, and Science. (2015). *Vision for a science nation. Responding to science, technology, engineering and mathematics: Australia's future.* Canberra, ACT: Commonwealth of Australia. Retrieved from https://www.voced.edu.au/content/ngv%3A78103.

Furtak, E., Seidel, T., Iverson, H., & Briggs, D. (2012). Experimental and quasi-experimental studies of inquiry-based science teaching. *Review of Educational Research, 82*(3), 795–813.

Han, H., & Lee, H. (2012). A study on the teachers' perceptions and needs of STEAM education. *Journal of Learner-Centred Curriculum and Instruction, 12*(3), 573–603.

Herro, D., & Quigley, C. (2017). Exploring teachers' perceptions of STEAM teaching through professional development: Implications for teacher educators. *Professional Development in Education, 43*(3), 416–438.

Hetland, L. (2013). Connecting creativity to understanding. *Educational Leadership, 70*(5), 65–70.

Hetland, L., Winner, E., Veenema, S., & Sheridan, K. M. (2007). *Studio thinking: The real benefits of visual arts education*. New York, NY: Teachers College Press.

Hong, S. M. (2016). A study on the employment status of science and technology personnel in science and technology and its implications[이공계 과학기술인력 고용 현황 분석과 시사점]. *Science & Technology Policy, 26*(3), 26–31.

Im, H.-J., & Lee, M.-H. (2012). *Pioneer professional development: STEAM pilot school case presentation* [2012 파이오니아 양성과정 연수 시범학교 사례발표 자료]. Busan, KR: Daeyeon Middle School. Retrieved from https://www.kofac.re.kr/upload/201204/1334539078854.pdf [in Korean].

Im, S.-M., Kim, Y., & Lee, T.-S. (2014). Analysis of elementary school teachers' perception on field application of STEAM education. *Journal of Science Education, 38*(1), 133–134.

International Baccalaureate Organization. (2005–2019). *Interdisciplinary learning*. Retrieved from https://www.ibo.org/programmes/middle-years-programme/curriculum/interdisciplinary/.

Jeong, S., & Kim, H. (2015). The effect of a climate change-monitoring program on students' knowledge and perceptions of STEAM education in the Republic of Korea. *Eurasia Journal of Mathematics, Science and Technology Education, 11*(6), 1321–1338.

Jon, J.-E., & Chung, H.-I. (2013). STEM report—Republic of Korea. In S. Marginson, R. Tytler, B. Freeman, & K. Roberts (Eds.), *STEM: Country comparisons* (pp. 33–46). Melbourne, AU: Report for the Australian Council of Learned Academies. Retrieved from https://acola.org/stem-country-comparisons-saf02/.

Koo, H. K., Chu, H.-E., Martin, S., & Choe, S. E. (2017, August 21–25). *Exploring the influence of students' science capital on scientific modelling process and conceptual understanding*. Poster presented at the meeting of the European Science Education Research Association, Dublin, Ireland.

Korea Foundation for the Advancement of Science and Creativity. (2012). *A study on the policy for promoting creative and convergent science and technology talents [창의 융합형과기인재 육성 정책 연구]*. Seoul, KR: Ministry of Education, Science, and Technology & Author.

Korea Foundation for the Advancement of Science and Creativity. (2018). *Final report on program development for STEAM education in 2018* [2018 년 융합 인재교육(STEAM)프로그램 개발 최종 보고서]. Seoul, KR: Author.

Korea Foundation for the Advancement of Science and Creativity. (2019). *Study on the monitoring and implementation of the 2015 science national curriculum* [2015 개정 과학과 교육과정 운영 모니터링 연구]. Seoul, KR: Author.

Korean Educational Development Institute. (2012). *Understanding of STEAM education through field application cases* (Issue Paper-OR 2012-02-02) *[현장 적용 사례를 통한 융합인재교육 (STEAM) 의 이해]*. Incheon, Kr: Author.

Lee, E-O. (2012). Possibilities and limitations of art education contents applied in STEAM. *Journal of Art Education, 33*, 287–314. Retrieved from http://www.kci.go.kr/kciportal/landing/article.kci?arti_id=ART001719754#nonen [in Korean].

Lee, J.-J., Jang, G.-C., & Han, S.-H. (1999). *Major tasks of Korea' s science and technology policy for the 21st century* [21 세기 과학기술정책의 부문별 과제]. Seoul, KR: Samsung Economic Research Institution.

Lee, J.-M., & Shin, Y.-J. (2014). An analysis of elementary school teachers' difficulties in the STEAM class. *Journal of Korean Elementary Science Education, 33*(3), 588–596.

Lee, K., Kim, K., & Lee, K.-J. (2013). An analysis of the lesson plans designed by teachers of the elementary STEAM leader schools. *Korean Education Review, 19*(2), 281–306.

Lim, S.-M., Kim, Y., & Lee, T.-S. (2014). Analysis of elementary school teachers' perception on field application of STEAM education. *Science Education Research Institute, 38*(1), 133–143.

Lim, Y.-N. (2012). Problems and ways to improve Korean STEAM education based on integrated curriculum. *Journal of Elementary Education, 25*(4), 53–80.

Liverpool Girls High School. (2019). *STEAM (science technology engineering arts mathematics).* Retrieved from https://liverpool-h.schools.nsw.gov.au/learning-at-our-school/steam–science-technology-engineering-arts-mathematics-.html.

Marginson, S., Tytler, R., Freeman, B., & Roberts, K. (2013). *STEM: Country comparisons. Report for the Australian Council of Learned Academie*s. Retrieved from https://acola.org/stem-country-comparisons-saf02/.

Martin, M. O., Mullis, I. V. S., Foy, P., & Stanco, G. M. (2012). TIMSS 2011 international results in science. Boston, MA: International Study Centre. Retrieved from https://files.eric.ed.gov/fulltext/ED544560.pdf.

Martin, M. O., Mullis, I. V. S., Gonzalez, E., Gregory, K., Smith, T., Chrostowski, S., Garden, R., & O'Connor, K. (2000). *TIMSS 1999 international science report: Findings from IEA's repeat of the third international mathematics and science study at the eighth grade.* Boston, MA: International Study Centre. Retrieved from https://timss.bc.edu/timss1999i/pdf/T99i_Math_All.pdf.

Miller, R. (Ed.). (1989). *Doing science: Images of science in science education.* Bristol, PA: Taylor & Francis.

Ministry of Education. (2015). *Primary and secondary schools' curriculum* (No. 2015-74). Seoul, KR: Author.

Ministry of Education, Science, and Technology. (2009). *The revised 2009 curriculum.* Seoul, KR: Author.

Ministry of Education, Science, and Technology. (2010). *Paving the way for Korea's future with creative talents and advanced science and technology (In Business Report 2011).* Seoul, KR: Author.

Ministry of Education, Science, and Technology. (2011a). *Science curriculum* (No. 2011-361). Seoul, KR: Author.

Ministry of Education, Science, and Technology. (2011b). *STEAM, the educational policy for 2011 year.* Seoul, KR: Author.

Morrison, J. (2006). *Attributes of STEM education: The student, the school, and the class-room.* Baltimore, MD: Teaching Institute for Excellence in STEM. Retrieved from https://www.partnersforpubliced.org/uploadedFiles/TeachingandLearning/Career_and_Technical_Education/Attributes%20of%20STEM%20Education%20with%20Cover%202%2.pdf.

New South Wales Education Standards Authority. (2018). *Science years 7-10 syllabus.* Sydney, AU: Author.

Noh, H. J., & Paik, S. H. (2014). STEAM experienced teachers' perception of STEAM in secondary education. *Journal of Learner-Centered Curriculum and Instruction, 14*(10), 375–402.

Norton, A., & Cakitaki, B. (2016). *Mapping Australian higher education.* Melbourne, AU: Grattan Institute.

Office of the Chief Scientist. (2014). *Science, technology, engineering, and mathematics: Australia's future.* Canberra, AU: Australian Government.

Park, H.-J., Baek, Y.-S., Sim, J., Son, Y.-A., Han, H., Byun, S.-Y., et al. (2014). *Enhancement of effectiveness of STEAM program and improvement of site [STEAM 프로그램 효과성 제고 및 현장 활용도 향상 기본연구].* Seoul, KR: KOFAC.

Park, H.-J., Byun, S.-Y., Sim, J., Baek, Y.-S., & Jeong, J.-S. (2016). A study on the current status of STEAM education. *Journal of the Korean Association for Science Education, 36*(4), 669–679.

Park, H.-J., Kim, Y.-M., Noh, S.-G., Lee, J.-O., Jeong, J.-S., Choi, Y.-H., et al. (2012). Components of 4C-STEAM education and a checklist for the instructional design. *Journal of Learner-Centered Curriculum and Instruction, 12*(4), 533–557.

Park, J., Chu, H.-E., & Martin, S. (2017, August 21–25). *Examining intercultural arts integrated STEM program.* Poster presented at the meeting of the European Science Education Research Association Dublin, Ireland.

Park, J.-H., & Shin, Y.-J. (2015). The effects of science-based STEAM class on the children's concept formation of heat transfer. *Journal of Science Education, 42*(2), 214–229.

Quigley, C. F., Herro, D., & Jamil, F. M. (2017). Developing a conceptual model of STEAM teaching practices. *School Science and Mathematics, 117*(1–2), 1–12.

Rabkin, N., & Hedberg, E. C. (2011). *Arts education in America: What the decline means for arts participation (Research Report #52)*. New York, NY: National Endowment for the Arts.

Rennie, L. J., Venville, G. J., & Wallace, J. (2012). *Knowledge that counts in a global community: Exploring the contribution of integrated curriculum*. London, UK: Routledge.

Saunders, W. L. (1992). The constructivist perspectives: Implications and teaching strategies for science. *School Science and Mathematics, 92*(3), 136–411.

Saunders-Stewart, K. S., Gyles, P. D. T., Shore, B. M., & Bracewell, R. J. (2015). Student outcomes in inquiry: Students' perspectives. *Learning Environments Research, 18,* 289–311.

Sawyer, R. K. (2006). *Explaining creativity: The science of human innovation*. Oxford, UK: Oxford University.

Sawyer, R. K., & DeZutter, S. (2009). Distributed creativity: How collective creations emerge from collaboration. *Psychology of Aesthetics, Creativity, and the Arts, 3*(2), 81–92.

Shin, H., Ahn, S., & Kim, Y. (2017). A policy analysis on the process-based evaluation: Focusing on middle school teachers in Seoul. *Journal of Curriculum and Evaluation, 20*(2), 135–162.

Shin, Y., & Han, S. (2011). A study of the elementary school teachers' perception in STEAM (science, technology, engineering, arts, mathematics) education. *Elementary Science Education, 30*(4), 514–523.

Sim, J., Lee, Y., & Kim, H.-K. (2015). Understanding STEM, STEAM education, and addressing the issues facing STEAM in the Korean context. *Journal of the Korean Association for Science Education, 35*(4), 709–723.

Singhai, P. (2017, September 12). Australia falling behind in science graduates and public funding. *Sydney Morning Herald.* Retrieved from https://www.smh.com.au/education/australia-falling-behind-in-science-graduates-and-public-funding-oecd-report-20170912-gyfigs.html.

Son, Y.-A., Jung, S.-I., Kwon, S.-K., Kim, H.-W., & Kim, D.-R. (2012). Analysis of prospective and in-service teachers' awareness of STEAM convergent education. *Institute for Humanities and Social Sciences, 13*(1), 255–284.

Song, S. S. (2004). Growth of science and technology activities in Korea and characteristics of society of scientists and engineers: An exploratory study. *Science & Technology Policy, 14*(1), 77–93.

Sullivan, G. (2006). Research arts in art practice. *Studies in Art Education, 48*(1), 19–35.

Taylor, P. C. (2016, August 7–9). *Why is a STEAM curriculum perspective crucial to the 21st century?* Paper presented at the Australian Council for Education Research, Brisbane: Australia. Retrieved from https://research.acer.edu.au/cgi/viewcontent.cgi?article=1299&context=research_conference.

Victorian Curriculum and Assessment Authority. (n.d.). *School-based assessment audit.* Retrieved from https://www.vcaa.vic.edu.au/administration/schooladministration/schoolbasedassessmentaudit.

Vygotsky, L. S. (1978). *Mind in society* (M. Cole, Trans.). Cambridge, MA: Harvard University Press.

Vygotsky, L. S. (1986). *Thought and language* (A. Kozulin, Trans.). Cambridge, MA: MIT Press.

Walker, C. L., & Shore, B. M. (2015). Understanding classroom roles in inquiry education: Linking role theory and social constructivism to the concept of role diversification. *SAGE Open, 5,* 4. https://doi.org/10.1177/2158244015607584.

Windschitl, M. (2004). Caught in the cycle of reproducing folk theories of "inquiry": How pre-service teachers continue the discourse and practices of an atheoretical scientific method. *Journal of Research in Science Teaching, 41*(5), 481–512.

Yakman, G. (2006). *STEM pedagogical commons for contextual learning* (Unpublished master's thesis). Virginia Polytechnic and State University, Blacksburg, Virginia, USA.

Yakman, G. (2008). *STE@M education: An overview of creating a model of integrative education.* Retrieved from https://www.iteea.org/File.aspx?id=86752&v=75ab076a.

Yakman, G., & Lee, H. (2012). Exploring the exemplary STEAM education in the US, as a practical education framework for Korea. *Journal of Korean Association for Science Education, 32*(6), 1072–1086.

Yoo, J., Hwang, S.-Y., & Han, I.-S. (2016). A comparative study of perceptions on STEAM education by the primary and secondary school teachers participated in the advanced STEAM teacher training program. *Journal of Research in Curriculum Instruction, 20*(1), 50–58.

Chapter 3
A Framework for Examining Teachers' Practical Knowledge for STEM Teaching

Kennedy Kam Ho Chan, Yi-Fen Yeh and Ying-Shao Hsu

3.1 Introduction

Around the world, there is an increasing call for providing K–12 students with quality science, technology, engineering, and mathematics (STEM) education to ensure that students will be able to engage and pursue STEM-related issues and careers (Metcalf, 2010; National Academy of Engineering & National Research Council, 2014). STEM education calls for new ways of teaching that go beyond the teaching of a particular discipline to teaching that involves an integration of different disciplines (Kelly & Knowles, 2016; Wang, Moore, Roehrig, & Park, 2011). Although what a teacher needs to know and be able to do in general for effective teaching and learning has been a subject of scholarly research (e.g., Cochran-Smith & Lytle, 1999; Guerriero, 2017; Verloop, van Driel, & Meijer, 2001), relatively less effort has been put into articulating the knowledge teachers need for effective STEM teaching (see exceptions: Allen, Webb, & Matthews, 2016; Saxton et al., 2014; Srikoom, Faikhamta, & Hanuscin, 2018). This leads to the central question: What knowledge does a teacher need for effective STEM teaching that leads to the valued student outcomes in STEM education? In this chapter, we pursue this question and propose a theoretical framework for examining and analyzing teachers' knowledge of STEM teaching. To achieve this goal, we first review the literature on STEM education to

Ying-Shao Hsu is a visiting professor at University of Johannesburg, South Africa.

K. K. H. Chan (✉)
Faculty of Education, The University of Hong Kong, Pok Fu Lam, Hong Kong
e-mail: kennedych@hku.hk; kennedyckh@gmail.com

Y.-F. Yeh
College of Teacher Education, National Taiwan Normal University,
Taipei, Taiwan

Y.-S. Hsu
Graduate Institute of Science Education, National Taiwan Normal University,
Taipei, Taiwan

© Springer Nature Singapore Pte Ltd. 2019
Y.-S. Hsu and Y.-F. Yeh (eds.), *Asia-Pacific STEM Teaching Practices*,
https://doi.org/10.1007/978-981-15-0768-7_3

identify STEM literacy and elements of effective STEM teaching. We then review the teacher knowledge literature to identify facets of knowledge needed for effective STEM teaching.

3.2 Integrated STEM Education

Around the globe, policy-makers, educators, industrial leaders, and business entrepreneurs have highlighted the critical importance of expanding and improving STEM education at the K–12 level. The call for STEM education goes beyond merely studying the four STEM subjects in isolated silos to tightening the connections within, between, and among these subjects in an integrated way that (a) reflects the nature of the work of most STEM professionals and (b) engages the interdisciplinary nature of most STEM issues. STEM education is advocated not only for workforce demands in science or engineering fields but also for the pursuit of informed citizenship: *STEM Literacy for All*. Unlike conventional approaches for developing talents in the science or engineering fields, STEM education focuses more on integrative learning experiences (Sanders, 2009) and soft skills development such as communication and teamwork (Hobbs, Clark, & Plant, 2018). It is worth pointing out that STEM should be viewed as a distinctive subject that is underpinned with some disciplinary features from each of the constituent disciplines. Yet, STEM is not a mere assembly of the four separate disciplines; rather, it should be viewed as a meta-discipline—a new discipline that is formed from the integration of other disciplines (Kennedy & Odell, 2014). As a meta-discipline, STEM is a cohesive entity that is greater than the sum of its parts, that is, the four respective disciplines.

3.2.1 STEM Literacy

STEM literacy can be conceptualized as comprising "the conceptual understanding and procedural skills and abilities for individuals to address STEM-related personal, social, and global issues" (Bybee, 2010, p. 31). Following PISA's framework for science, reading, and mathematical literacy, Bybee proposed that STEM competencies include three aspects, namely, identifying STEM issues, explaining issues from STEM perspectives, and using STEM information. These competencies reflect features of STEM projects like context-dependent, practice-based (i.e., meaningful implication of knowledge and skills in practices), creativity pursued as well as both the disciplinary knowledge and generic thinking abilities involved. STEM-related issues can be real-life situations or problems to solve, which explains why STEM literacy should be viewed as educational outcomes most students should achieve.

A meta-level STEM literacy is also worthy of pursuit, especially when real-life situations and problems are so complicated that there is often no single or easy solution. Viewed from this perspective, STEM literacy should not be merely viewed as a composite of *S*, *T*, *E*, and *M* literacies. Rather, core competence should entail learners developing literacy in terms of how problem-solvers activate what has been learned from various disciplines and then create feasible solutions in the context of problem-solving. Yeh, Hsu, Wu, Yang, and Lin (under review) proposed five competencies that are critical to problem-solving but are primitively incubated in separate disciplines. These competencies are analogical reasoning, contextualization, quantitative thinking, prediction, and reflective ability. Taking contextualization as an example, problem solvers first need to decontextualize problems into what is familiar such as processing calculation. The solution prototypes then need to be recontextualized using the right languages for the audience targeted for the follow-up mass production or marketing. Contextualization and decontextualization can involve problem definition and rationale expression of design in the engineering field (Atman et al., 2007). However, it should be noted that transfer of learning is not easily or automatically achieved (Dixon & Brown, 2012; Johnson, Dixon, Daugherty, & Lawanto, 2011). It is believed that these meta-level competencies can be greatly nurtured in STEM-related or problem-solving tasks if they can be purposefully emphasized with disciplinary connections.

Zollman (2012) added nuances to the idea of STEM literacy when he urged that three domains of STEM literacy be strengthened, under the ultimate goal of "STEM literacy for learning [rather than] learning for STEM literacy" (p. 12). Apart from knowledge and skills to address STEM issues, he contends that reflection helps learners improve their solutions as well as become quicker and better thinkers for any new challenges in the future. The ability to self-regulate determines how efficient students may be in strategic problem-solving (e.g., making plans or collaborating with others), which allows them to better understand themselves and build their self-identity from exploration. Finally, students' STEM literacy elaborates the stages of thinking of actions, linking between movements, and automatically refining performance. The three domains (i.e., cognition, affection, and psychomotor) contribute to the ultimate objectives for STEM education: *STEM literacy for continual learning* (Zollman, 2012). Therefore, STEM literacy encompasses not only knowledge and skills requisite for problem-solving but also a set of generic skills and learning dispositions that enable life-long learning. Students who attain and keep enhancing their meta-disciplinary STEM literacy will be the successful candidates to fulfill the STEM pipeline demands and future careers.

In summary, STEM literacy builds on *S*, *T*, *E*, and *M* literacies—but what matters the most is how students use and integrate their related knowledge and competencies from the respective disciplines adequately and flexibly to solve problems encountered or create products to satisfy needs. To develop such an interdisciplinary (or transdisciplinary) literacy demands not just knowledge or competency development, it is critical that students develop interdisciplinary ways of thinking as well as persistent but sustainable ways of learning.

3.2.2   Effective STEM Teaching

Vasquez, Sneider, and Comer (2013) identified a continuum of levels of STEM integration in terms of interconnection between the respective STEM disciplines. *Multidisciplinary* involves learning the core concepts and skills separately in each discipline but situating them in a common theme. *Interdisciplinary* entails learning closely-linked concepts and skills from two or more disciplines for deepening the learning of those concepts and skills. *Transdisciplinary* involves the application of concepts or skills from more than two disciplines to real-life problems or projects. Despite the varied perspectives of STEM integration, there appear to be common-alities in effective STEM teaching. First, effective STEM teaching should have an explicit focus on content integration across the disciplines (e.g., Ring, Dare, Crotty, & Roehrig, 2017). Second, effective STEM teaching should not only focus on the development of content knowledge but also foster skills development such as innovative problem-solving and inquiry skills (e.g., Wang et al., 2011). It logically follows that effective STEM teaching foregrounds the use of student-centered pedagogies such as inquiry and problem-based learning approaches (e.g., Breiner, Harkness, Johnson, & Koehler, 2012; Sanders, 2009) and the use of real-life contexts (Breiner et al., 2012). Moore, Johnson, Peters-Burton, and Guzey (2015) developed a STEM integration framework for effective teaching that succinctly identifies six essential elements: a personally meaningful, motivating, and engaging context; engineering design challenges; learning from failure through redesign; embedding mathematics and/or science content; use of student-centered pedagogies; and an emphasis on teamwork and communication.

The above review suggests that effective STEM teaching demands that teachers teach in a completely new way from traditional, teacher-directed, content teaching. What teacher knowledge is required to support this new way of teaching? The following sections address this question by first reviewing teacher knowledge literature and then proposing the nature and composition of knowledge required for effective STEM teaching.

3.2.3   Teacher Knowledge for Effective Teaching

What teachers need to know for effective teaching has attracted scholarly attention for many decades (e.g., Cochran-Smith & Lytle, 1999; Shulman, 1986; Verloop et al., 2001). We define teacher knowledge as the sum of knowledge a teacher possesses that guides his/her actions (Carter, 1990). A teacher may consciously or unconsciously use or refrain from using some of his/her knowledge of teaching.

Shulman (1986, 1987) proposed that teacher professional knowledge is comprised of seven categories: content knowledge, general pedagogical knowledge, curriculum knowledge, pedagogical content knowledge (PCK), knowledge of learners and their characteristics, knowledge of educational contexts, and knowledge of educational ends, purposes, and values. Shulman's work is influential for at least two reasons. First, it reinforces the notion that teachers are professionals with a unique province of professional knowledge that is not shared by others (i.e., content specialists). Second, it highlights that teachers' knowledge comprises not only knowledge that is generic in nature (i.e., applicable to different subject domains) but also the knowledge that is specific to the teaching of a particular body of content.

Although Shulman's ideas were well received, many debates about the nature and composition of teacher knowledge continue to exist in the field (e.g., Chan & Hume, 2019). A group of researchers working in the area of science teacher knowledge met in 2012 to propose a consensus model for teacher professional knowledge and skills (Gess-Newsome, 2015). The model (Fig. 3.1) makes explicit several characteristics of teacher professional knowledge. First, this model differentiates two different facets of teacher knowledge: general knowledge bases for teaching and topic-specific professional knowledge. The former is generic across topics and includes knowledge categories such as assessment knowledge, pedagogical knowledge, etc. The

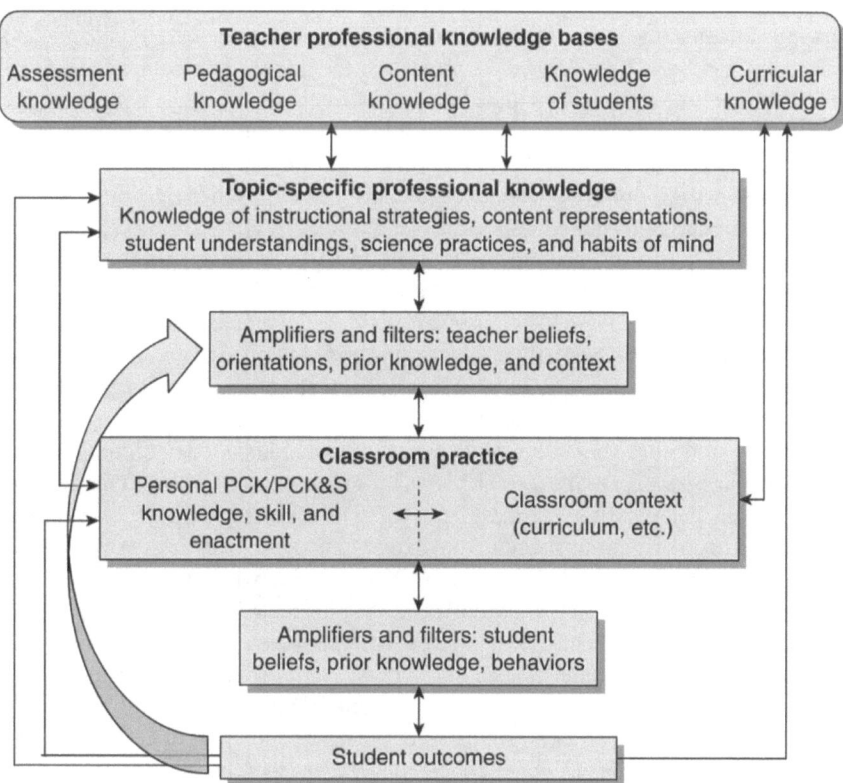

Fig. 3.1 Adapted from "A Model of Teacher Professional Knowledge and Skill including PCK" by J. Gess-Newsome, 2015, in *Reexamining pedagogical content knowledge in science education* (p. 31). Copyright 2015 by Routledge Publishing

latter includes topic-specific knowledge for teaching a particular topic. Second, the model distinguishes between canonical and personal knowledge. Canonical knowledge is generated by research or best practice, which can have a normative function while personal knowledge is private and idiosyncratic in nature, which is developed from a teacher's classroom experience. Cochran-Smith and Lytle (1999) indicated that canonical knowledge can be regarded as knowledge *for* practice whereas personal knowledge is knowledge *of* the practice. Moreover, the consensus model highlights that a teacher's classroom practices are informed by topic-specific professional knowledge and general knowledge bases for teaching and that the teacher's beliefs serve as a filter or amplifier that mediates the translation of knowledge into classroom practices.

3.3 Teacher Knowledge for Effective STEM Teaching

We see effective STEM teaching as comprising a set of teaching practices (e.g., engaging students in motivating contexts, use of student-centered pedagogies) informed by teachers' knowledge. We assert that the knowledge required for effective STEM teaching is broad and multifaceted. Different types of teacher knowledge integrate to inform a teacher's decisions for planning, enactment, and reflection on his/her STEM instruction. In other words, teachers' knowledge informs teachers' planning, real-time monitoring, and adjustment as well as post hoc reflection. Teachers' beliefs about STEM integration (e.g., Wang et al., 2011), for example, can serve as a filter or amplifier to mediate the translation of knowledge into the teachers' practices.

We are interested in using the consensus model for characterizing teachers' personal knowledge, which we call teachers' practical knowledge for STEM teaching. Teachers' practical knowledge is personal, context-bound, and guides teachers' action in concrete and specific situations (van Driel, Beijaard, & Verloop, 2001). Therefore, teachers are generators of their own practical knowledge through reflection on their classroom practices—STEM teaching is supported by both topic-specific and generic teacher knowledge. Like others (e.g., Davis & Krajcik, 2005), we believe that teachers need discipline-specific knowledge to be able to "help students understand the authentic activities of a discipline, the ways knowledge is developed in a particular field, and the beliefs that represent a sophisticated understanding of how the field works" (p. 5). Hence, teachers need different types of knowledge that may be topic-specific (i.e., specific to teaching a particular concept), discipline-specific (i.e., specific to teaching STEM discipline or a particular *S*, *T*, *E*, *M* discipline), or domain-general (i.e., general knowledge about teaching) for effective STEM teaching. Although some scholars have conceptualized teacher knowledge for STEM teaching as STEM PCK or PCK for STEM (e.g., Allen et al., 2016; Saxton et al., 2014; Srikoom et al., 2018), we are reluctant to using this label because we believe that knowledge required for effective STEM teaching embraces (a) some elements that are related to student skills development (e.g., teaching of problem-solving skills) and (b) other elements that are content-specific (e.g., teaching of disciplinary content). We also argue that STEM teaching goes beyond aiming at merely teaching students a particular body of content to concepts from different disciplines and their interconnections. Hence, we prefer the term *Practical Knowledge for STEM Teaching*.

To conceptualize the composition of teachers' knowledge for teaching STEM, we drew on the consensus model and prior work (e.g., Allen et al., 2016, Magnusson et al., 1999; Saxton et al., 2014). Our analysis suggests that, apart from content knowledge, there are four important knowledge components, namely, assessment, pedagogy, curriculum, and students (Fig. 3.2). These knowledge components may be topic-specific, domain-specific, or domain-general in nature.

We envisage that the quality of teachers' knowledge also differs as a result of several factors, such as teachers' years of STEM teaching experience and their formal education. Expert teachers are characterized by a rich and elaborated knowledge base (Borko & Livingston, 1989; Leinhardt & Greeno, 1986). Moreover, experts are known to have a flexible knowledge base that allows rapid retrieval of knowledge for teaching performance. Their knowledge goes beyond *knowing that* to *knowing how* and *knowing why*. In other words, experts do not adhere to context-free rules but are

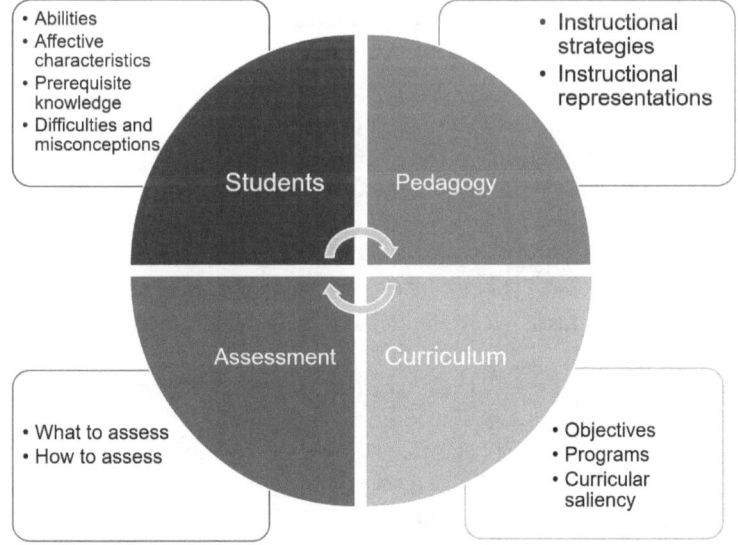

Fig. 3.2 Teacher knowledge for effective STEM teaching

able to apply the principles in practices based on the situations (Dreyfus & Dreyfus, 1986). As such, expert STEM teachers would not only have knowledge for teaching STEM that is greater in quantity but also of higher quality. The knowledge is more detailed, more contextualized, and situated in different teaching cases and real-life teaching examples.

To summarize, our conceptualization of teachers' practical knowledge for effective STEM teaching (details in Fig. 3.2) takes into account the major components of teacher knowledge and acknowledges that the knowledge may exist in varying degrees of specificity (i.e., topic-specific, domain-specific, generic), quantity, and quality (i.e., concreteness). The four components are knowledge about assessment, knowledge of pedagogy, knowledge about curricula, and knowledge about students. It is noteworthy that the components of knowledge serve only for analytic purposes. In reality, the boundaries between knowledge components are fuzzy and teachers draw on these knowledge components as a whole in an integrated fashion to inform their planning, enactment, and reflection on their STEM instruction. Teachers' beliefs (e.g., their beliefs about STEM integration) may mediate the translation of their knowledge into actual classroom practices.

3.4 Interview Protocols for Investigating Teacher Knowledge

A variety of tools and strategies have been used to investigate teachers' professional knowledge (e.g., Baxter & Lederman, 1999; Black & Halliwell, 2000). Data collection instruments include questionnaires, surveys, interviews, and classroom observations (e.g., Chan & Hume, 2019; van Driel, Berry, & Meirink, 2015). Each instrument has its own unique affordances and limitations. Acknowledging the inherent challenges of measuring teachers' cognition (Kagan, 1990), we propose, as a starting point, the use of semistructured interviews to elicit teachers' knowledge for teaching STEM.

The interview protocol (Table 3.1) is structured into three sections. The first part elicits teachers' conceptions about STEM education as well as their beliefs about the purposes of STEM education. We believe that how a teacher conceptualizes STEM education greatly influences their STEM instruction. Based on the consensus model, we see teachers' beliefs as an important amplifier and filter in mediating teachers' use of knowledge. The second part probes the teachers' knowledge for STEM teaching in terms of the four teacher knowledge components (i.e., curriculum, assessment, students, and pedagogy). The questions specifically prompt teachers to differentiate between teaching that focuses only on disciplinary content from STEM teaching that entails not only content but also interconnections between/amongst concepts and skills from different disciplines. Teachers are asked to provide examples to illustrate their ideas as far as possible. This provides a window into how they draw on their knowledge to design curriculum and tailor instructions in classrooms. The third

Table 3.1 Interview protocol to elicit a teacher's knowledge for STEM teaching

Part 1: Teacher's views about STEM education and purposes of STEM education
1. What are the first words or phrases that come to your mind when you hear the word "STEM"?
2. How do you define "STEM"?
3. Why do you want to implement STEM education?
4. What do you think are the important elements for STEM literacy?

Part 2: Teacher's knowledge for teaching STEM
5. (a) What learning objectives or goals do you set for your students in STEM education?
(b) What learning objectives or goals do you think your students set for themselves?
6. To achieve the goals, how do you design and implement your courses?
7. What learning difficulties do you think your students have about STEM?
8. How can you know whether your students have achieved the learning objectives in STEM?
9. What are the challenges that you encounter when you do STEM education? How do you deal with them?
10. What do you think are the differences when you teach disciplinary content and STEM courses?
(a) Do you use different strategies?
(b) How do students respond to these two types of instruction?
(c) Do you use different assessments?
(d) How is the curriculum different?

Part 3: Teacher's professional development experience related to STEM
11. What professional development do you think STEM teachers need to be equipped with?
12. How do you build up your professional learning in STEM education?
13. Have you attended any teacher learning communities? Do you think it helpful in equipping yourself to teach STEM?
14. Is there anything that you think the government, the schools, or the university can do to improve the quality of STEM education?

part of the interview examines the teachers' professional development experiences related to STEM. The questions prompt teachers to reflect on their professional development experiences to identify perceived needs in their future professional development. Such information may be useful for professional developers to design powerful learning environments to promote teachers' STEM teaching.

It is hoped that through analyses of the voices, stories, and examples shared by teachers with varying STEM teaching experience, we will be able to elicit, capture, and document the critical knowledge for STEM teaching. Specifically, we would like to characterize the nature and content of knowledge for STEM teaching and identify patterns among teachers that surpass the idiosyncratic level o f individual stories and narratives.

3.5 Conclusion

This chapter has engaged the question of what teacher knowledge is requisite for effective STEM teaching. We approached this question by reviewing the learning outcomes (i.e., STEM literacy) advocated in STEM education. We identified several key elements of effective STEM teaching and theorized the knowledge that supports practices conducive to effective STEM teaching through a review of the STEM education and teacher knowledge literature. We ended by explicating the design of an interview protocol that serves as a tool to elicit teacher knowledge for STEM teaching. The proposed teacher knowledge framework can serve as a useful analytic tool for researchers to characterize the nature and content of teacher knowledge that informs effective STEM teaching. The interview protocol will also reveal the professional development needs of teachers for effective STEM teaching from the voices of the teachers. The chapters that follow will exemplify the findings based on an empirical investigation of teachers using these tools in different Asian countries.

References

Allen, M., Webb, A. W., & Matthews, C. E. (2016). Adaptive teaching in STEM: Characteristics for effectiveness. *Theory Into Practice, 55*(3), 217–224.

Atman, C. J., Adams, R. S., Cardella, M. E., Turns, J., Mosborg, S., & Saleem, J. (2007). Engineering design processes: A comparison of students and expert practitioners. *Journal of Engineering Education, 96*(4), 359–379.

Baxter, J. A., & Lederman, N. G. (1999). Assessment and measurement of pedagogical content knowledge. In J. Gess-Newsome & N. G. Lederman (Eds.), *Examining pedagogical content knowledge: The construct and its implications for science education* (pp. 147–161). Dordrecht, The Netherlands: Kluwer.

Black, A. L., & Halliwell, G. (2000). Accessing practical knowledge: How? Why? *Teaching and Teacher Education, 16,* 103–115.

Borko, H., & Livingston, C. (1989). Cognition and improvisation: Differences in mathematics instruction by expert and novice teachers. *American Educational Research Journal, 26*(4), 473–498.

Breiner, J. M., Harkness, S. S., Johnson, C. C., & Koehler, C. M. (2012). What is STEM? A discussion about conceptions of STEM in education and partnerships. *School Science and Mathematics, 112*(1), 3–11.

Bybee, R. W. (2010). Advancing STEM education: A 2020 vision. *Technology and Engineering Teacher, 70*(1), 30–35.

Carter, K. (1990). Teachers' knowledge and learning to teach. In W. R. Houston (Ed.), *Handbook of research on teacher education* (pp. 291–310). New York, NY: Macmillan.

Chan, K. K. H., & Hume, A. (2019). Towards a consensus model: Literature review of how science teachers' pedagogical content knowledge is investigated in empirical studies. In A. Hume, R. Cooper, & A. Borowski (Eds.), *Repositioning pedagogical content knowledge in teachers' knowledge for science teaching* (pp. 3–76). Singapore: Springer.

Cochran-Smith, M., & Lytle, S. L. (1999). Relationships of knowledge and practice: Teacher learning in communities. In A. Iran-Nejad & P. D. Pearson (Eds.), *Review of research in education* (Vol. 24, pp. 249–305). Washington, DC: American Educational Research Association.

Davis, E., & Krajcik, J. (2005). Designing educative curriculum materials to promote teacher learning. *Educational Researcher, 34*(3), 3–14.

Dixon, R., & Brown, R. A. (2012). Transfer of learning: Connecting concepts during problem solving. *Journal of Technology Education, 24*(1), 2–17.

Dreyfus, H., & Dreyfus, S. (1986). *Mind over machine: The power of human intuition and expertise in the era of the computer.* New York, NY: The Free Press.

Gess-Newsome, J. (2015). A model of teacher professional knowledge and skill including PCK: Results of the thinking from the PCK Summit. In A. Berry, P. J. Friedrichsen, & J. Loughran

(Eds.), *Re-examining pedagogical content knowledge in science education* (pp. 28–42). New York, NY: Routledge.

Guerriero, S. (2017). *Pedagogical knowledge and the changing nature of the teaching profession.* Paris, France: OECD Publishing.

Hobbs, L., Clark, J. C., & Plant, B. (2018). Successful students – STEM program: Teacher learning through a multifaceted vision for STEM education. In R. Jorgensen & K. Larkin (Eds.), *STEM education in the junior secondary: The state of play* (pp. 133–168). Singapore: Springer.

Johnson, S. D., Dixon, R., Daugherty, J., & Lawanto, O. (2011). General versus specific intellectual competencies: The question of learning transfer. In M. Barak & M. Hacker (Eds.), *Fostering human development through engineering and technology education* (pp. 55–74). Rotterdam, The Netherlands: Sense.

Kagan, D. M. (1990). Ways of evaluating teacher cognition: Inferences concerning the Goldilocks principle. *Review of Educational Research, 60*(3), 419–469.

Kelley, T. R., & Knowles, J. G. (2016). A conceptual framework for integrated STEM education. *International Journal of STEM Education, 3*(1), 11.

Kennedy, T., & Odell, M. (2014). Engaging students in STEM education. *Science Education International, 25*(3), 246–258.

Leinhardt, G., & Greeno, J. (1986). The cognitive skill of teaching. *Journal of Educational Psychology, 78,* 75–95.

Magnusson, S., Krajcik, J., & Borko, H. (1999). Nature, sources, and development of pedagogical content knowledge for science teaching. In J. Gess-Newsome & N. G. Lederman (Eds.), *Examining pedagogical content knowledge: The construct and its implications for science education* (pp. 95–132). Dordrecht, The Netherlands: Kluwer Academic.

Metcalf, H. (2010). Stuck in the pipeline: A critical review of STEM workforce literature. *InterActions: UCLA Journal of Education and Information Studies, 6*(2), Article 4, 1–20.

Moore, T. J., Johnson, C. C., Peters-Burton, E. E., & Guzey, S. S. (2015). The need for a STEM road map. In C. C. Johnson, E. E. Peters-Burton, & T. J. Moore (Eds.), *STEM road map: A framework for integrated STEM education* (pp. 3–12). New York, NY: Routledge.

National Academy of Engineering & National Research Council. (2014). *STEM Integration in K-12 education: Status, prospects, and an agenda for research* (M. Honey, G. Pearson, & H. Schweingruber, Eds.). Washington, DC: National Academies Press. https://doi.org/10.17226/18612.

Ring, E. A., Dare, E. A., Crotty, E. A., & Roehrig, G. H. (2017). The evolution of teacher conceptions of STEM education throughout an intensive professional development experience. *Journal of Science Teacher Education, 28*(5), 444–467.

Sanders, M. (2009). STEM, STEM education, STEMmania: A series of circumstances has once more created an opportunity for technology educators to develop and implement new integrative approaches to STEM education championed by STEM education reform doctrine over the past two decades. *The Technology Teacher, 68*(4), 20–26.

Saxton, E., Burns, R., Holveck, S., Kelley, S., Prince, D., Rigelman, N., et al. (2014). A common measurement system for K–12 STEM education: Adopting an educational evaluation methodology that elevates theoretical foundations and systems thinking. *Studies in Educational Evaluation, 40,* 18–35.

Shulman, L. S. (1986). Those who understand: Knowledge growth in teaching. *Educational Researcher, 15*(2), 4–14.

Shulman, L. S. (1987). Knowledge and teaching: Foundations of the new reform. *Harvard Educational Review, 57*(1), 1–22.

Srikoom, W., Faikhamta, C., & Hanuscin, D. L. (2018). Dimensions of effective STEM integrated teaching practice. *K-12 STEM Education, 4*(2), 313–330.

van Driel, J. H., Beijaard, D., & Verloop, N. (2001). Professional development and reform in science education: The role of teachers' practical knowledge. *Journal of Research in Science Teaching, 38*(2), 137–158.

van Driel, J. H., Berry, A., & Meirink, J. A. (2015). Research on science teacher knowledge. In N. G. Lederman & S. K. Abell (Eds.), *Handbook of research on science education* (Vol. 2, pp. 848–870). New York, NY: Routledge.

Vasquez, J., Sneider, C., & Comer, M. (2013). *STEM lesson essentials, grades 3–8: Integrating science, technology, engineering, and mathematics*. Portsmouth, NH: Heinemann.

Verloop, N., van Driel, J. H., & Meijer, P. (2001). Teacher knowledge and the knowledge base of teaching. *International Journal of Educational Research, 35*(5), 441–461.

Wang, H.-H., Moore, T. J., Roehrig, G. H., & Park, M. S. (2011). STEM integration: Teacher perceptions and practice. *Journal of Pre-College Engineering Education Research, 1*(2), 1–13. https://doi.org/10.5703/1288284314636.

Yeh, Y.-F., Hsu, Y.-S., Wu, H.-K., Yang, K.-L., & Lin, K.-Y. (under review). Problem solving in STEM education—From discipline-based to integrative design.

Zollman, A. (2012). Learning for STEM literacy: STEM literacy for learning. *School Science and Mathematics, 112*(1), 12–19.

Chapter 4
Instructional Knowledge of STEM: The Voices of STEM Teachers in Taiwan

Yi-Fen Yeh and Ying-Shao Hsu

4.1 Background

Our changing world is one in which education is ever increasingly important. We need our future citizens to be ready for forthcoming challenges; therefore, contemporary education goals must focus on literacy (e.g., scientific literacy) and the development of survival skills (e.g., critical thinking, adaptability) that empower students to work, solve problems, and strive to be lifelong learners (Wagner, 2008). STEM—science, technology, engineering, and mathematics—education emphasizes interdisciplinary knowledge and skill development, higher order thinking through problem-solving, and connections between schooling and the world. As such, it is an area that attracts educators from a wide variety of academic subjects.

Interdisciplinary education is one of the foremost challenges for today's teachers. Students have reported that they learn far better from interdisciplinary teaching and learning rather than when a multidisciplinary pedagogy is employed (Jones, 2009). Teachers commonly receive a discipline-specific education via their academic majors, but those teachers who seldom engage in authentic inquiry may be less adaptable and receptive to interdisciplinary teaching and learning designs. Teaching interdisciplinary topics demands not only that teachers become proficient in related fields through self-learning and collaboration with other educators but also that they cultivate abilities like systemic and cross-linked thinking (Burandt & Barth, 2010). Teacher qualification is one priority that must be considered if we are to launch

---

Ying-Shao Hsu is a visiting professor at University of Johannesburg, South Africa.

Y.-F. Yeh (✉)
College of Teacher Education, National Taiwan Normal University, Taipei, Taiwan
e-mail: yyf521@ntnu.edu.tw

Y.-S. Hsu
Graduate Institute of Science Education, National Taiwan Normal University, Taipei, Taiwan
e-mail: yshsu@ntnu.edu.tw

© Springer Nature Singapore Pte Ltd. 2019
Y.-S. Hsu and Y.-F. Yeh (eds.), *Asia-Pacific STEM Teaching Practices*,
https://doi.org/10.1007/978-981-15-0768-7_4

and sustain quality STEM education. Since various school levels (i.e., primary, middle, and high) have recruited teachers of science, mathematics, and technology but not necessarily engineering, we are interested in learning more about how teachers develop STEM curricula and how well they have developed their STEM instructional knowledge.

4.2 Developing Teachers for STEM Education

The emergence of STEM education is a response to the needs of a twenty-first-century workforce that can support cutting-edge industry development and citizens who can apply what they learned in school to solve life's problems (Caprile, Palmen, Sanz, & Dante, 2015; Charette, 2013). STEM literacy, the comprehensive goal of STEM education, is a composite construct of the knowledge and abilities that individuals rely on to address complex issues involved with various component topics (Bybee, 2013). Therefore, preparation of STEM teachers must first clarify the ultimate goals for STEM learners. Only then will we have a better understanding of (a) the knowledge and skills with which STEM teachers must be equipped and (b) how that acquisition can be facilitated.

4.2.1 Development of Instructional Knowledge for STEM

STEM education is an integrative approach that involves science, technology, engineering, and mathematics and also serves as a broader conceptual space not strictly limited to these four disciplines. Other possible areas may include disciplines such as the environment, economics, and medicine and creative artistic endeavors (Tarnoff, 2011; Zollman, 2011). Furthermore, scholars have suggested that STEM's greatest value is the purposeful integration of these disciplines into solving real-world problems (Labov, Reid, & Yamamoto, 2010). Bybee (2010) viewed STEM as an integrative subject where discipline-specific ways of thinking are combined and promoted, such as the identification of STEM issues, explanation of topics from STEM perspectives, and use of STEM information to solve problems. The literacy that a STEM education develops should also loop back to encourage lifelong learning effectiveness by strengthening learners' cognition (e.g., reflective abstraction), affection (e.g., self-regulation), and psychomotor skills (e.g., being an automatic learner) (Zollman, 2011). We take STEM literacy to be a metadisciplinary collection of knowledge, skills, and attitudes. Considering that STEM knowledge is usually topic based or interdisciplinary in nature, it may not be practical to expect the future workforce to develop full expertise in STEM for every unforeseen issue.

To help students develop this interdisciplinary literacy, teachers must be equipped with the skills of a particular profession and its related pedagogical skills. Previous literature has indicated that effective STEM teaching relies on content integration, a personal ability to solve problems innovatively or by conducting authentic inquiry, instruction in problem-solving, inquiry based in student-centered approaches, and the use of real-life contexts (Breiner, Harkness, Johnson, & Koehler, 2012; Chan, Yeh, & Hsu, under review; Ring, Dare, Crotty, & Roehrig, 2017; Wang, Moore, Roehrig, & Park, 2011). However, considering that teachers traditionally develop their instructional knowledge as domain or topic specific and even as personally developed, it can be quite difficult to reach a consensus regarding the definition of STEM pedagogical content knowledge (PCK). Chan et al. (under review) proposed a generic framework for practical instructional knowledge that is composed of four key domains: knowledge of assessment, pedagogy, curriculum, and students. Teachers effective in teaching STEM must develop related instructional ideas and experiences of varying degrees of specificity (i.e., domain specific and generic) and quality (i.e., quantity and concreteness).

Curriculum integration is ideal because knowledge that is relevant outside of schools lacks defined disciplinary boundaries. However, this holistic nature is not necessarily so when multiple disciplines are integrated in a contrived fashion (Beane, 1995). STEM education can be a useful solution encouraging cohesive integration and meaningful learning if "it encompasses real world, problem-based learning that links the disciplines through cohesive and active teaching and learning approaches" (English, 2016, p. 2). Developing such a curriculum demands that teachers either be knowledgeable about the topics being taught or able to communicate and work successfully with colleagues possessing different areas of expertise. Knowledge gaps may even appear among teachers of closely integrated disciplines such as science and mathematics (Stinson, Harkness, Meyer, & Stallworth, 2009).

Inequitable representations of the four STEM subjects involved are quite common in the literature. Often, curricula are science dominated or only two of the four disciplinary categories are emphasized, demonstrating the challenges in designing STEM curricula and opportunities for instructional knowledge development (Vasquez, Sneider, & Comer, 2013). How to encourage teachers to step out of their comfort zone to develop integrated curricula and sustain their professional development is critical to STEM teacher development, especially when topics are usually inquiry based and real-life contextualized. Teachers have to be motivated and willing to explore the target issues beyond the traditional disciplinary boundaries to develop curricula and instructional guidance that align with students' learning needs.

4.2.2 Teacher Community for Teacher Development

Discipline-specific teachers can engage in leading interdisciplinary courses like STEM by collaborating with educators of different subject specialization or by engaging in self-learning focused on interdisciplinary issues. Professional learning communities (PLCs), which are often self-initiated organizations, have become an excellent resource for teachers seeking to care for and learn from one another, embrace a vision beyond the scope of individual members, overcome difficulties in instruction, and induce change in practice and belief (Lambert, 2003; Tam, 2015). Attributes of successful PLCs include (a) being oriented toward and striving for better student learning progress; (b) continuously working together to find better teaching practices, enhance personal learning development, and encourage school improvement; and (c) aiming for evidence-based learning progress and seeking out areas of improvement (DuFour, 2004). However, traditional professional development (PD) efforts are focused on the needs of content area specialists; therefore, many PLCs are discipline based. In contrast, PLCs for STEM education must be interdisciplinary and, thus, should focus more on how teachers from different content areas and perspectives can be attracted from the greater community and learn to communicate and work well with other professions.

Project Lead the Way (PLTW), a large nonprofit organization, offers middle and high school STEM education programs (https://www.pltw.org/). Brophy, Klein, Portsmore, and Rogers (2008) argued that PLTW courses should engage students in support topics (e.g., the scientific process, engineering problem-solving, applications of technology), cross-disciplinary subjects (e.g., understanding how technology works with other tools, using mathematical knowledge to solve nonmathematical problems), and soft skills (e.g., effective communication, working with others). STEM teachers, whether they adopt an embedded or integrated approach (Roberts & Cantu, 2012), must step out of their respective comfort zones, teach beyond their familiar boundaries, and solve the problems that emerge during this process. Avery and Reeve (2013) recommended that PD for STEM teachers should include providing exemplar engineering design challenges, strengthening teachers' understanding of curriculum standards, learning evaluation methods for students' group performances, developing teachers' STEM PCK as integral to the profession through STEM lesson design, and engaging STEM concepts in instructional materials. Reynolds, Yazdani, and Manzur (2013) found that effective PD for STEM should engage teachers in accomplishing engineering-based research and projects on their own. Inclusion of these features would enhance the design of PLCs for STEM educators.

4.3   Method

The central focus of this chapter is documenting how subject-specific teachers developed STEM curricula and their level of instructional knowledge about STEM. Yin (2003) suggested that case studies are explanatory, exploratory, and descriptive in nature and are appropriate where a phenomenon within a real-life context or after an intervention can be investigated in order to reveal how and why the events occur or variables interact. We selected and documented the case of a PLC that had developed a series of STEM curricula and interviewed the teachers who comprised its membership.

4.3.1 Background of the Case

There were several reasons for choosing these teachers to serve as the PLC case study. First, they had spent at least 1 year developing thematic STEM-related courses as electives for Grades 11 and 12; two teachers had engaged even longer in related curriculum planning and implementation. Moreover, this PLC was formed to develop a STEM-related curriculum for the High-Scope III Project which aimed to "integrate emerging S&T [science and technology] of everyday life into their curriculum for fostering development of innovative S&T" (Ministry of Education, 2018, p. 9). The High-Scope Project is led by the Taiwan Ministry of Education, which offers grants to encourage middle schools, high schools, and colleges to develop topic-specific curricula for several disciplines or areas of study. It should be noted that the high school in this study had several PLCs; however, the others were not focused on STEM curriculum development.

The six participating teachers interviewed in the STEM PLC studied were all males and their backgrounds included physics, mathematics, technology, and the arts. They were teaching in a girls' high school (Grades 10–12) in the southern part of Taiwan. Recently, compulsory education in Taiwan has been extended to 12 years. Ninth graders in middle school have several paths to the high school they would like to attend. One path is to take the scholastic academic examination and achieve the required score for the desired school and the other path is a school-based selection process. Since the case study school was one of the oldest schools in that city and well known for its high-level student performance, the students in this school generally had a high aptitude for academics and most were likely to attend good universities.

4.3.2 Data Collection and Analysis

We used an interview protocol designed to reveal and document teachers' instructional knowledge of STEM categorized into four knowledge domains: curricula, students, instructional strategies, and assessment (Chan et al., under review). Each knowledge domain had two to three indicators (Table 4.1) from which the interview questions were developed (see Table 3.1 in Chap. 3).

Table 4.1 Codebook for instructional knowledge of STEM

Knowledge of curriculum (KC)	Knowledge of students (KS)	Knowledge of pedagogy (KP)	Knowledge of assessment (KA)
• Curriculum goals • Programs and materials • Identification of salient ideas	• Student abilities • Affective characteristics • Prerequisite knowledge • Difficulties or misconceptions	• Instructional representations • Instructional strategies	• What to assess • How to assess

Each interview lasted 30–60 min, prior to which all interviewees completed a background survey related to academic degrees and teaching experiences. All interview data were transcribed and coded through NVivo (https://www.qsrinternational.com/nvivo/home). The interview data were coded using the smallest meaningful episodes that were judged by knowledge indicators and the explicit (either general or topical) examples they included. Teachers' proficiency in each knowledge subset was evaluated using a consistent system where teachers would receive 1 point for a general example or a topical example. Scores of each knowledge category were composed of a score of general knowledge and a score of topical knowledge. These scores were calculated based on the overall response level they achieved for the indicators within a category over the numbers of corresponding indicators. That is, if it was a four-indicator category like KS, the accumulated score that a teacher might earn for KS-topical would be 4 at most if she gave many topical examples for each of the indicators within the KS. Likewise, the maximum score for the KS-general knowledge would be 4 if she provided 1 or more examples for each indicator. For fair comparisons across categories with different numbers of indicators, the accumulated score would be divided by the number of indicators in that category to provide a category average of general and topical knowledge: The KS-topical would need to be averaged by its four indicators. For example, a teacher would receive 1 point for a topical example within the indicator KP-1 (instructional representations) regardless of the frequency of specific examples she offered. If she did not offer any topical examples of KP-2 (instructional strategies), her final category score for KP-topical would be adjusted to 0.5. The comparisons of teachers' STEM knowledge in this study were made on the basis of knowledge categories, instead of indicators. Furthermore, grouping teacher knowledge into STEM-general or STEM-specific knowledge based on experience examples would enrich our discussions of teachers' instructional knowledge development.

4.4 Findings

We report the PLC profile from the aspects of the six teachers' backgrounds and brief curriculum descriptions, their performance in the four knowledge domains, and the cross-referenced PD and PLC needs.

4.4.1 Backgrounds of Teachers and Their Curricula

The case PLC was a school-based, curriculum development community, sharing the comprehensive goal of developing a series of STEM-related courses. The backgrounds of the six teachers being interviewed (e.g., subjects for which they were responsible, years of teaching experience) are shown in Table 4.2. When surveyed about their confidence in STEM teaching, all but T 5 felt confident teaching STEM-related courses. T1 was the leader of the PLC and a key person for bringing in external support (e.g., writing grant proposals to support curriculum development, bringing in PD support from the university). He won a teaching award in 2018 from the Ministry of Education, which is viewed as the highest honor for teachers in Taiwan. The average amount of teaching experience for these teachers was 13 years, with a range of 5–21 years.

The PLC had developed a total of eight courses within the four themes (Fig. 4.1). Six of these eight courses were open to students who were interested in creating products or making things, whether they were oriented toward science or the liberal arts. Only two courses were more advanced and were offered solely to students who had chosen the science-oriented academic track. Taking the quadcopter course as an example, the teacher introduced aviation principles and related physical concepts, followed by engaging students to play with paper airplanes and simulate aviation in mobile phone apps. Students were later guided to manipulate DC electric motors through programming with Arduino (https://www.arduino.cc/), use 3D modeling to make simulated bamboo rafts, and control four DC motors' revolutions per minute (rpms) using both Arduino and cell phone apps. These activities allowed students to engage with concepts like transistors in technology, lift and angle of aviation in physics, rpms of DC motors used for quadcopter aviation, etc. After these introductory sessions, students physically experienced the application of inertia detectors on Segways® and small quadcopters, and applied acceleration and angular velocity to improve their respective quadcopter. Each group needed to remotely control the quadcopters, fly them through pathways in balanced aviation, and land them at an appointed location. Last, students visited the aerospace department at the local university to learn about different quadcopter applications in real life and gain experience controlling aerial drones. Their learning was evaluated via worksheets and tests and they also created an aerial photograph exhibition at the annual school celebration.

4.4.2 Teachers' Performance of STEM Instructional Knowledge

We evaluated the six teachers' STEM instructional knowledge based on their interview responses (Interview questions can be found in Chap. 3, this book). Their responses regarding the four knowledge domains are illustrated by type (i.e., general and specific knowledge) in Fig. 4.2. Certain patterns were identified from these teachers' performance.

Table 4.2 Teachers' subject backgrounds and STEM teaching experiences

	T1	T2	T3	T4	T5	T6
Academic degree	Ph.D.	Master's	Master's	Master's	Master's	Master's
Subject area (years of experience)	Mathematics (21)	Physics (5)	Information and computer education (11)	Arts (9)	Technology (19)	Mathematics (13)
STEM courses (years of experience)	Mathematical modeling, cloud computing, sleep control (2)	Arduino circuit implementation, program physics (1)	Game physics (1)	Incredible mechanisms (1)	Robotic sensing and control, theory of flight dynamics, control of quadcopters (5)	Artificial intelligence, trademark design (3)
STEM-related project experiences	High-scope III	High-scope III, STEM curriculum development	Game physics curriculum development in high-scope III	School actualization, mobile learning, high-scope III	High-scope III, school actualization, homogenized curriculum development	Artificial intelligence curriculum development for high-scope III
Confidence in offering STEM courses	Confident regarding innovations in technology and tools, aligning with educational reforms, teacher education, etc.	Fair level of confidence and is learning to do better	Self-assured, due to connections with personal discipline profession and from working within the PLC	Self-assured because "STEM plus Arts" is fun for students	Could do better when more experiences are gained	Self-assured after receiving good responses to leadership of science fairs and STEM teaching experiences
Notes	Teacher Award in 2018 (national level)					Frequently invited STEM education lecturer

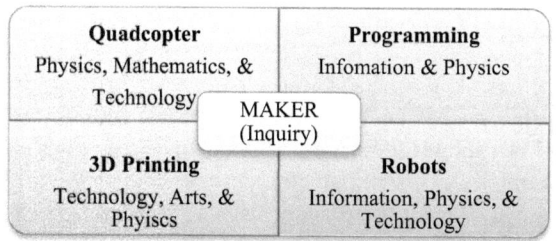

Fig. 4.1 The four themes of STEM-related courses developed by the professional learning community

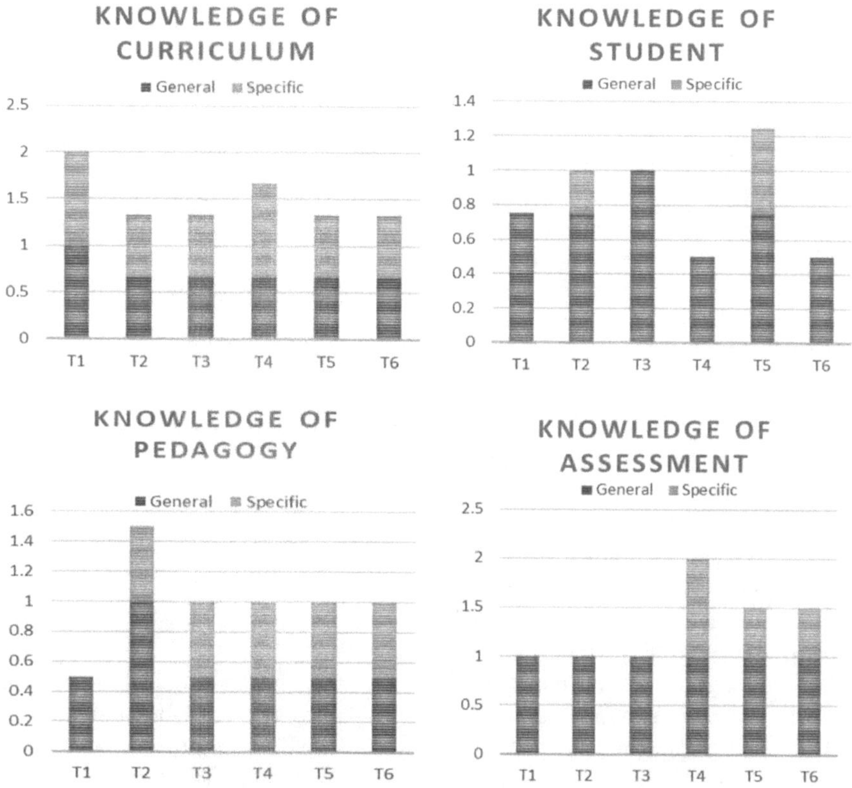

KNOWLEDGE OF CURRICULUM

KNOWLEDGE OF STUDENT

KNOWLEDGE OF PEDAGOGY

KNOWLEDGE OF ASSESSMENT

Fig. 4.2 Teachers' interview performance in STEM instructional knowledge. Note that the vertical scales are different across the four knowledge components

4.4.2.1 Knowledge of Curriculum

Among the four domains, teachers' knowledge of curriculum (KC) seemed to be better developed in terms of both general and specific knowledge. However, we attribute the high-level knowledge scores to the teachers' responses about curriculum goals. They were the designers and practitioners of the courses, so they were well suited to explicate what they expected students to achieve, not only with general (G) but also specific (S) topical examples.

> T2: I think students should have the ability to analyze problems and engage in self-learning, since what we can teach them is very limited. They raise more questions under our appropriate guidance and take an active role in learning from those questions. … We cannot complement that part of knowledge. [KC-1G]

> T3: For this semester, my ultimate goal for students was to let them construct works of kinetic art. That is, throughout the history of the arts there has been art education, and artwork like this already exists. It can be found in sculpture. Gear theory has been used to make sculptures move, but they also have their own aesthetic value and paradigms. Then I started to look for … a physics teacher to collaborate with, hoping that he could help me take care of the structural mechanism of gears in physics. Students reviewed what they should have known and then tried to connect these two things. Therefore, I explicitly told my students that now that I'm an art teacher, in this course we will eventually get into art. Many sciences are involved in our art course. [KC-1S]

> T1: There could be many sources for projects. For example, I could design a brewer for coffee cups. The taste of the coffee is determined by the temperature, the warmth it keeps, and its water flow. [KC-3S]

4.4.2.2 Knowledge of Students and Pedagogy

Knowledge of student (KS) was the teachers' weakest knowledge domain, especially at the level of specific knowledge. T5 was the teacher with the highest total scores from the four domains; furthermore, T5 had developed his practical knowledge about specific issues across all domains. T2 and T4 were the second highest. The quote from T2 below illustrates his command of instructional knowledge about accommodating student learning needs with abundant topical knowledge that was transformed from his physics background and integrated with engineering applications. Based on the teachers' background survey, their experience with conducting projects focused on STEM curriculum development was a powerful indicator of their proficiency level in STEM instructional knowledge (Table 4.2).

> T2: Definitely we need to guide students, without a doubt. For this question [KP-2G], they came up with many sources for lift, but they only knew that it's lifting [KS-4S]. Then I guided them, such as by asking, "Why does it lift when the propellers rotate?" They thought that it could be wind. Then I continued, asking, "Then why do airplanes and fixed-wing airplanes lift up without propellers?" They answered, "They had wings." Then I used an analogy. "Helicopters have no wings, only propellers, so what's the correlation between the propellers and wings of planes?" [KP-2S]. Based on this guidance, they captured the idea that propellers are like the wings of planes. Planes have wings but they're fixed, while

[the rotating] propellers interact with wind [air]. Forces come from these interactions. So, I prompted them to consider whether the wind [air movement] was one reason contributing to lift [KP-2S]. They then started to draw a force diagram of wings and wind [air movement] and analyzed the pushing power [KP-2S]. After that, I asked them if the angles of the propellers were related or if angles could be another variable. [KP-2S]

4.4.2.3 Knowledge of Students and Assessment

Almost all of the teachers were experienced in teaching their respective subjects (i.e., the component STEM subjects) but were still developing their knowledge about and experience with STEM teaching. Therefore, it made sense that they had better levels of performance in terms of general knowledge. By comparison, knowledge of students (KS) and knowledge of assessment (KA) are types of instructional knowledge usually developed later along the PCK development continuum. KC and KP are more teacher-initiated anticipatory domains, while KS demands more reflective teaching experiences (e.g., noticing students' needs and cognitive development). Assessment is usually where teachers reflect and develop after they have the goals and instructional activities of the curricula designed.

> T5: Assessments are the other difficulties for me. In terms of assessment … first, when it comes to groups, it makes me wonder how to … should I give the whole group the same score? Or how do I give different scores to 2 students in the same group? Of course their scores were ultimately based on the final results and their worksheets. And sometimes students may take official leaves and I don't feel like I can deduct their scores for that. I'm still looking for a proper way to deal with that. But generally speaking, my assessments were formative ones, mainly worksheets, since I don't want to give them any more academic pressure. [KA-2G]. … For students with special performance, like being my assistant in class or willing to answer questions frequently, I would offer them extra points. But I don't deduct students' points since I don't know them really well. [KA-1G]

4.4.3 Teachers' Perspectives Toward PLC

Since the PLC had been operating for 1–2 years for the purpose of curriculum development, information regarding how the teachers felt about the PLC offers valuable information for planning PD. Three strands were identified from their interview data: operation of the PLC, collaborative teaching, and self-learning (Table 4.3).

The PLC case could be viewed as a success if the shared goals were achieved, the participants grew in terms of their teaching, and a good working atmosphere was established among the participating teachers. A culture of co-learning and co-teaching also contributes to PLC's success. These six teachers prioritized the importance of self-learning and self-motivation, implying that they respected the other participants and were mutually supportive of each other. STEM classrooms that are open to interested faculty are friendly environments that accommodate both teachers

Table 4.3 Cross-referenced comments regarding critical aspects of a professional learning community

	PLC	Collaborative teaching	Self-learning
T1 (M)	• Fun to work with colleagues, which can be useful for future interdisciplinary learning	• We set courses as electives; I also invite teachers from different disciplines to co-teach courses	• STEM teachers are highly motivated in this area • As an academics-oriented school, teachers here have strong academic knowledge, which enables them to design interesting courses or activities. It would be another story if they were not motivated
T2 (S)	• Technology teachers help solve problems through practicality • Most courses are co-taught by two teachers, so we do our best to brainstorm different potential projects and enrich course content	• These courses are highly demanding with continuous problem-solving, so two teachers are needed. I may spend 10 min guiding a student who has a problem	• The two courses I offer are not closely related to my academic area (physics), so I spend an extensive amount of time self-learning • The most distinctive feature of STEM courses is teachers' self-learning rather than student learning. We all feel that we get back to the era of being students. We learn knowledge, solve problems, and build up something sufficient. It's quite fun during the process
T3 (T)	• We teachers in the PLC learn from each other and may consult previous course instructors for instructional ideas	• Better to prepare courses with colleagues from different fields since teachers will definitely face problems outside their profession	• Teachers should be open-minded and like to learn and try

(continued)

Table 4.3 (continued)

	PLC	Collaborative teaching	Self-learning
T4 (A)	• My colleagues like to help me when I find something I am not good at • Our current PLC works well because we are very close	• Our office is near the technology teacher's office, and we are good friends • T5 is a teacher full of educational idealism and he invites me to co-teach with him • I also work with T6, and we used GeoGebra to guide students to design logos	• STEM teachers should be interested in making things • T1 likes to build robots with LEGO®
T5 (T)	• Easier to learn from our physics and math teachers when we encounter problems in those areas	• We work with the professors and graduate students in the university's Maker Center Graduate students serve as tutors in our classes	• The Department of Aeronautics and Astronautics offered us a one-semester course to take
T6 (M)	• We are still developing courses for Grade 12. We've learned how others deliver courses and what concepts are engaged, which helps me to design courses	• I've co-taught with two other teachers. I became a student when the technology teacher was teaching programming. … Now the technology teacher is teaching with the art teacher (T3). T3 wants me to guide students to learn related math knowledge and then graphically design trade logos on computers. So I go to T3's classroom and learn with the students while T3 goes to mine	• STEM teachers are usually self-motivated, since everything starts with taking action

and students learning with one another. It is also important that school authorities support the co-teaching and co-planning system and establish related policies (e.g., by reasonably sharing teaching hours, arranging curriculum development hours, etc.). Successful co-teaching would require the course instructor to be the lead planner, especially when weekly topics and assessments are mutually conceptualized and negotiated as well as the teachers' expertise needs to be properly engaged (Chanmugam & Gerlach, 2013).

However, it should be noted that PD may not bring all positive impacts. There are barriers (e.g., administrative constraints, interpersonal issues, logistical or scheduling issues) or tensions attending PLCs (e.g., work pressure, shared learning, intrapersonal growth) in addition to the benefits of interdisciplinary PD (e.g., professional growth, enhanced trust and respect for colleagues, shared responsibility and collaborative problem-solving, collaborative research and co-teaching opportunities) (Miller & Stayton, 1998; Schaap et al., 2018). Both co-learning and co-teaching should be pursued, especially for interdisciplinary courses like STEM, since STEM knowledge and abilities are not only topic based but also demand flexibility in order to solve various problems in the process.

4.5 Final Remarks

This chapter discussed how STEM teachers who were subject specialists originally developed their instructional knowledge of STEM and how their PLC shaped their instructional knowledge and enabled them to develop a series of STEM courses. STEM education encourages students to complete projects or solve problems with different levels of complexity and difficulty. Teachers should consider if adequate difficulties are embedded in the STEM projects or problems since these life- and job-related tasks offer students good opportunities to deepen their knowledge, abilities, higher order thinking, and even expanding career options. This is why researchers have suggested that successful STEM education depends on teachers' pedagogy rather than content (Tytler, Osborne, Williams, Tytler, & Clarke, 2008) and that curriculum resources and assessments must be well planned to align with student needs and program goals (Kennedy & Odell, 2014). The teachers in this study may have developed good PCK in the specific subjects they teach, but they still need time to reflect on their STEM teaching experiences in order to transform their discipline-specific PCK into STEM PCK.

PD and PLC are critical to continuous teacher development—although the focus and format may be slightly different for teachers of single and interdisciplinary subjects. In the case examined here, experienced teachers united for the purpose of developing a series of STEM curricula. They knew their students and curriculum standards, so they preferred to develop motivating and inspiring courses. Drawing upon their mature PCK, they were enthusiastic about attending college-level engineering courses for a semester, learning with their fellow PLC participants, and collaborating in course design and teaching. These teachers ensured that they had prepared themselves until they were fully ready to deliver the designed STEM courses and able to guide their students. The leader of this PLC was important since he strategically brought in necessary resources (e.g., a topic-specific PD course offered by the university, curriculum development grants) and encouraged teachers from different fields to explore different topics and develop curricula collaboratively. To accommodate the many possibilities in STEM education, a sustainable PLC should have members with talents from different fields who are striving for a shared goal with full support from other stakeholders.

Acknowledgements This research is supported by the Institute for Research Excellence in Learning Sciences and the Higher Education Sprout Project of National Taiwan Normal University and sponsored by the Ministry of Education in Taiwan.

References

Avery, Z. K., & Reeve, E. M. (2013). Developing effective STEM professional development programs. *Journal of Technology Education, 25*(1), 55–69.

Beane, J. A. (1995). Curriculum integration and the disciplines of knowledge. *Phi Delta Kappan, 76*(8), 616–622.

Breiner, J. M., Harkness, S. S., Johnson, C. C., & Koehler, C. M. (2012). What is STEM? A discussion about conceptions of STEM in education and partnerships. *School Science and Mathematics, 112*(1), 3–11.

Brophy, S., Klein, S., Portsmore, M., & Rogers, C. (2008). Advancing engineering education in P-12 classrooms. *Journal of Engineering Education, 97*(3), 369–387.

Burandt, S., & Barth, M. (2010). Learning settings to face climate change. *Journal of Cleaner Production, 18,* 659–665.

Bybee, R. W. (2010). Advancing STEM education: A 2020 vision. *Technology and Engineering Teacher, 70*(1), 30–35.

Bybee, R. W. (2013). *The case for STEM education: Challenges and opportunities*. Arlington, VA: NSTA Press.

Caprile, M., Palmén, R., Sanz, P., & Dente, G. (2015). *Encouraging STEM studies for the labour market (Report for European Parliament's Committee on Employment & Social Affairs)*. Brussels, Belgium: European Union.

Chan, K., Yeh, Y.-F., & Hsu, Y.-S. (under review). A framework for examining teachers' practical knowledge for STEM teaching. In Y.-S. Hsu (Ed.), *Asia-Pacific STEM teaching practices: From theoretical frameworks to practices*. Singapore: Springer.

Chanmugam, A., & Gerlach, B. (2013). A co-teaching model for developing future educators' teaching effectiveness. *International Journal of Teaching and Learning in Higher Education, 25*(1), 110–117.

Charette, R. N. (2013, August 30). The STEM crisis is a myth. *IEEE Spectrum.* Retrieved from http://spectrum.ieee.org/at-work/education/the-stem-crisis-is-a-myth.

DuFour, R. (2004). What is a professional learning community? *Education Leadership, 61*(8), 6–11. Retrieved from http://www.ascd.org/publications/educational-leadership/may04/vol61/num08/What-Is-a-Professional-Learning-Community%C2%A2.aspx.

English, L. D. (2016). STEM education K-12: Perspectives on integration. *International Journal of STEM Education, 3*(3). https://doi.org/10.1186/s40594-016-0036-1.

Jones, C. (2009). Interdisciplinary approach—Advantages, disadvantages, and the future benefits of interdisciplinary studies. *ESSAI, 7,* Article 26. https://dc.cod.edu/essai/vol7/iss1/26/.

Kennedy, T. J., & Odell, M. R. L. (2014). Engaging students in STEM education. *Science Education International, 25*(3), 246–258.

Labov, J. B., Reid, A. H., & Yamamoto, K. R. (2010). Integrated biology and undergraduate science education: A new biology education for the twenty-first century? *CBE Life Sciences Education, 9*(1), 10–16.

Lambert, L. (2003). *Leadership capacity for lasting school improvement*. Alexandria, VA: Association for Supervision and Curriculum Development.

Miller, P. S., & Stayton, V. D. (1998). Blended interdisciplinary teacher preparation in early education and intervention: A national study. *Topics in Early Childhood Special Education, 18*(1), 49–58.

Ministry of Science and Technology. (2018). *Ministry of science and technology: Republic of China (Taiwan)*. Taipei, Taiwan: Author. Retrieved from https://www.most.gov.tw/most_ebook/en/.

Reynolds, D., Yazdani, N., & Manzur, T. (2013). STEM high school teaching enhancement through collaborative engineering research on extreme winds. *Journal of STEM Education: Innovations and Research, 14*(1), 12–19.

Ring, E. A., Dare, E. A., Crotty, E. A., & Roehrig, G. H. (2017). The evolution of teacher conceptions of STEM education throughout an intensive professional development experience. *Journal of Science Teacher Education, 28*(5), 444–467.

Roberts, A., & Cantu, D. (2012). Applying STEM instructional strategies to design and technology curriculum. In *Proceedings of the Technology Education in the 21st Century*. PATT 26 Conference, Stockholm, Sweden, 111–118.

Schaap, H., Louws, M., Meirink, J., Oolbekkink-Marchand, H., Want, A. V. D., & Meijer, P. (2018). Tensions experienced by teachers when participating in a professional community. *Professional Development in Education*. https://doi.org/10.1080/19415257.2018.1547781.

Stinson, K., Harkness, S. S., Meyer, H., & Stallworth, J. (2009). Mathematics and science integration: Models and characterizations. *School Science and Mathematics, 109*(3), 153–161.

Tam, A. C. F. (2015). The role of a professional learning community in teacher change: A perspective from beliefs and practices. *Teachers and Teaching, 21*(1), 22–43.

Tarnoff, J. (2011, May 25). STEM to STEAM—Recognizing the value of creative skill in the competitiveness debate. *The Huffington Post*. Retrieved from http://www.huffingtonpost.com/john-tarnoff/stem-to-steam-recognizing_b_756519.html.

Tytler, R., Osborne, J., Williams, G, Tytler, K., & Clarke, J. C. (2008). *Opening up pathways: Engagement in STEM across the primary-secondary school transition. A review of the literature concerning supports and barriers to science, technology, engineering and mathematics engagement at primary-secondary transition*. Melbourne, Australia: Deakin University.

Vasquez, J. A., Sneider, C., & Comer, M. (2013). *STEM lesson essentials, grades 3–8: Integrating science, technology, engineering, and mathematics*. Portsmouth, NH: Heinemann.

Wagner, T. (2008). Rigor defined. *Educational Leadership, 66*(2), 20–24.

Wang, H.-H., Moore, T. J., Roehrig, G. H., & Park, M. S. (2011). STEM integration: Teacher perceptions and practice. *Journal of Pre-college Engineering Education Research, 1*(2), 1–13.

Yin, R. (2003). *Case study research: Design and methods*. Thousand Oaks, CA: Sage.

Zollman, A. (2011). Learning for STEM literacy: STEM literacy for learning. *School Science and Mathematics, 112*(1), 12–19.

Chapter 5
Teachers' Conceptions About STEM and Their Practical Knowledge for STEM Teaching in Hong Kong

Valerie W. Y. Yip and Kennedy Kam Ho Chan

5.1 Introduction

STEM education has been one of the major reforms in the fields of science, technology, and mathematics education since the 1990s. It aims to support students to better acquire transferrable skills such as problem-solving, critical thinking, and collaboration (Morrison, 2006). Students who are proficient in STEM would be more capable of (a) apply knowledge and skills in solving real-life problems and (b) engage in professional dialogues of STEM-related issues (Bybee, 2010).

The importance of STEM education in Hong Kong was highlighted in the Chief Executive's 2015 policy address (HKSAR Government, 2015), which resulted in a policy document by the Education Bureau (2016) on incorporating STEM education formally into the school curricula through the Science, Technology, and Mathematics Key Learning Areas. The Government also provided one-off financial support to primary and secondary schools to implement their school-based STEM programs. Since then a number of projects and professional development programs have proliferated to support teachers to implement the new curricula.

Given the inter- or even transdisciplinary nature of STEM, ideal STEM education should be integrative to resemble out-of-school learning and the professional life of scientists or engineers. Despite this priority directive, previous research has consistently identified that meaningful connections between the four STEM disciplines are missing in the curricula (e.g., Dickerson, Cantu, Hathcock, McConnell, & Levin, 2016). Teachers play a gatekeeper role in what happens in the classrooms; therefore, it is critical that teachers possess informed conceptions and knowledge related to integrating STEM into the existing curricula and/or developing new curricula. This chapter aims to examine Hong Kong teachers' conceptions of integrated STEM education and their knowledge of STEM education by focusing on an in-service teacher

V. W. Y. Yip (✉) · K. K. H. Chan
Faculty of Education, The University of Hong Kong, Pok Fu Lam, Hong Kong
e-mail: valyip@hku.hk

© Springer Nature Singapore Pte Ltd. 2019
Y.-S. Hsu and Y.-F. Yeh (eds.), *Asia-Pacific STEM Teaching Practices*,
https://doi.org/10.1007/978-981-15-0768-7_5

and a preservice teacher who are at different stages of developing their understanding of STEM education. These findings can inform professional development efforts necessary for STEM teachers to meet the instructional challenges in this reform.

5.1.1 Integrated STEM Education

The Next Generation Science Standards (National Research Council [NRC], 2013) has put the core ideas, epistemic practices, and interdisciplinarity alongside science teaching. Science teachers are required to teach science, engineering, and mathematics in an integrated manner. Within the literature, there is evidence suggesting that mathematics and science integration has proven to have positive influences on students' achievement (e.g., Hurley, 2001).

Effective integrated STEM instruction has been examined by a number of studies (e.g., Berlin & White, 1995; Honey, Pearson, & Schweingruber, 2014; Kennedy & Odell, 2014; Stohlmann, Moore, & Roehrig, 2012; Zemelman, Daniels, & Hyde, 2005). A synthesis of the literature elicits eight recommended practices informed by previous research. Some of these practices can be common to general teaching (e.g., items 1 and 5), while the others can be more related to integrated STEM instruction (e.g., items 2 and 3).

1. Understand students' knowledge, capabilities, and learning difficulties;
2. Include technology and engineering into science and mathematics curricula if separate STEM courses are not available;
3. Make STEM integration explicit to students, for instance, highlight the knowledge and practices of a particular discipline to students;
4. Focus on big ideas, concepts, processes, and representations and their connections;
5. Support students to develop disciplinary knowledge to facilitate their application in integrative contexts;
6. Use a problem-solving approach to situate students in real-life challenges;
7. Facilitate active inquiry, engineering design, reasoning, argumentation, reflection, and collaboration; and
8. Make use of assessment effectively, for example, use multiple strategies and different points of data collection to inform future instruction.

By adopting the recommended practices sensibly, students can develop STEM literacy that is not simply adding the four literacy strands of science, technology, engineering, and mathematics together (Zollman, 2012). With effective STEM teaching, these four literacies can be integrated to facilitate students' learning in the cognitive, psychomotor, and affective domains and hence meet their personal, economic, and social needs.

5.1.2 Challenges of Integrated STEM Instruction

Despite the importance of integrating the four disciplines in teaching, STEM education from K–12 has been criticized for lacking integration (English, 2016). This might be explained by two reasons.

First, it has been difficult to define what STEM should encompass because of its interdisciplinary nature, which is unique to the discipline-specific traditions that historically drive teacher education. There are fundamental differences in the purposes of studying science and engineering: science aims to account for natural phenomena and engineering targets to find solutions for daily-life problems. Developing a good and holistic understanding of *STEM* can be challenging. For example, technology may be regarded as the mere application of science (National Academy of Engineering [NAE] & NRC, 2002), and science comes before technology. Many people also mistakenly assume that technology emphasizes the products as *nouns* while a few people would recognize it as human innovation in the *making process* (Yore, 2011). Such conceptual understandings may require a lot of effort to overcome. In addition, educators have been told the diversified interpretations of the acronym *STEM*, ranging from simply technical applications (e.g., Keefe, 2010) to more social, cultural, economic, and political applications that reflect real-world situations (Zeidler, 2016). These broad differences might influence the pedagogical decisions of STEM teachers; for instance, can the application of technology, which has been quite common in science and mathematics classrooms, be regarded as STEM teaching? Is it necessary for students to discuss the social impacts of any STEM innovations all the time? More importantly, what perspectives do teachers have about STEM?

The second reason for a lack of integration lies within the teacher workforce. While discussing the need of purposeful design and inquiry in teaching STEM, Sanders (2009) put forward the following scenario:

> Many technology teachers are *fond of* saying they teach science and math[ematics] in their technology education programs. In truth, it is *exceedingly rare* for a technology teacher to *explicitly identify* a specific science or mathematics concept or process *as a desired learning outcome* and even rare for technology teachers to *assess* a science or mathematics learning outcomes. (p. 21)

This scenario can be applicable to science and mathematics teachers when they teach STEM. While teachers appreciate the value of integration (i.e., fond of), their practices *rarely* demonstrate it. Teachers are usually educated as subject specialists. The education background of individuals (Kennedy & Odell, 2014) and the initial teacher education that teachers receive usually focuses on discrete disciplines (Blackley & Howell, 2015). Although scientific inquiry and the design of technological artifacts have been implemented in science classes, it is still difficult for teachers who have deeply rooted beliefs and practices in teaching particular subjects to move beyond their comfort zone and adopt instructional strategies they might be less certain about (items 1–8 in the previous section).

The two major challenges of integrated STEM education informed by the literature include teachers' (a) diffused concepts of what STEM entails and (b) lack of understanding about how teaching multiple disciplines at once would be possible. Since Hong Kong is new to this educational reform, the following three questions arise.

1. What conceptions about STEM do Hong Kong teachers hold?
2. What practical knowledge do Hong Kong teachers possess for STEM teaching?
3. How can teachers be supported to teach STEM effectively?

By examining the Hong Kong teachers' conceptions and knowledge about STEM at different stages of developing an understanding of STEM instruction, this study attempts to shed light on how to better support Hong Kong teachers to develop effective STEM teaching.

5.2   Methods

This qualitative research (Merriam, 2015) is part of a larger study that aimed to understand Hong Kong teachers' conceptions and practical knowledge for teaching STEM. We focused on how the background and learning experiences of the participants influenced their conceptions on STEM and their knowledge bases. Purposeful sampling (Patton, 2002) was employed to identify (a) in-service teachers who were responsible for the planning, implementation, and evaluation of their school-based STEM curricula and who already had fundamental training in and at least 2 years' teaching experience with STEM, and (b) preservice teachers who had completed a 24-h undergraduate, science-related pedagogical course on STEM teaching. This could ensure that the novices, although without much practical teaching experience, had a basic understanding of STEM education.

The participants completed a background questionnaire and joined semi-structured interviews (~1.5 h). The purposes of the interviews were three-folded as follows: (a) to understand teachers' conceptions about STEM teaching; (b) to explore their practical knowledge for STEM teaching; for which the knowledge bases could be divided into pedagogical knowledge, curricular knowledge, student knowledge, and assessment knowledge (Gess-Newsome, 2015); and (c) the participants' professional development experience and their professional development needs (see Chap. 3, this book). The interviews were transcribed verbatim and analyzed by basic qualitative techniques (Patton, 2002). First, we identified the relevant parts of the interview reflective of the teachers' conceptions of STEM. Codes were developed to capture the nature of the teachers' conceptions. Second, we identified their knowledge of STEM teaching by analyzing the interview transcripts to identify the four dimensions within teacher professional knowledge bases. To ensure the trustworthiness of the findings, we employed investigator triangulation (Denzin, 1989). Interpretations were thoroughly discussed to ensure that an accurate account is presented.

5.2.1 The Two Cases

Most of the participants, particularly the in-service teachers, had different STEM-related majors in their academic studies. In this chapter, we focused on a subset of the larger dataset that is representative of science teachers. The two case teachers reported in this study primarily had more understanding of science and science education. Wendy was a secondary school teacher with 8 years' teaching experience. She had taught biology, chemistry, and junior secondary science and was the head of the school's biology department. Working in a school with socially deprived students, Wendy took up the team leader role to develop a junior secondary STEM curriculum that used Maker—a technology-based DIY (do it yourself) strategy that emphasizes making and creating—to support students in becoming more proficient in generic skills, taking more responsibility for self-directed learning, and being more confident in their learning. Within the lessons, the students formed groups to develop products they found necessary to support the socially disadvantaged groups such as physically disabled people. Although Wendy had the experience to teach science, she had joined many STEM-related teacher professional development programs as STEM had been new to her, especially the technology and engineering. At the time of the study, she was a member of a STEM teachers' network and had collaborated with several external agents (e.g., local Maker laboratories such as Making of Loft) to facilitate her work as a STEM education leader in her school.

Charles was a preservice physics teacher who was studying in an undergraduate double-degree program that mainly focused on science teaching and learning at the time of the study. He reported in his questionnaire that he had not had many opportunities to learn about STEM teaching before joining the core pedagogical course. The course professor indicated that Charles' understandings of STEM education improved substantially after the 24-h course. During the lessons, Charles demonstrated great interest in the activities.

5.3 Findings

In this section, we briefly discuss the two case teachers' conceptions about STEM. Then, we discuss the content and nature of their knowledge for teaching STEM. We highlight their similarities and differences in the conceptions about STEM as we present data about their knowledge. Our analysis is based on the excerpts of the interviews, and the italicized texts represent the key ideas expressed in the interview quotes.

5.3.1 Teachers' Conceptions About STEM

Both Wendy and Charles had been educated as science teachers. Setting against this background, both put science as the most important component in STEM. Charles stated:

> Everyone can have different interpretations of STEM. *As science teachers, we should focus on [teaching] science.* ... On the other hand, we need to have some understanding about technology, engineering, and mathematics so that we know with whom we can collaborate in planning STEM-related lessons or activities.

Similarly, Wendy emphasized science as the prerequisite for STEM. Without the scientific principles, the other components of STEM might become less meaningful. She stated:

> As science teacher, I always put "S" [for science] at the first place. *STEM becomes complete with S.* For example, to make a vehicle move, engineers need to build different parts. The building [process] has to be backed up by scientific principles.

By saying "should focus on science" and "becomes complete with S," both teachers showed how they interpreted STEM. Science is used to support technology and engineering, and STEM education aims to facilitate students to better understand how scientific concepts are applied to develop a STEM product. It is worth pointing out that the teachers might hold a common misconception that technologies must be produced with an understanding of the underlying science concepts (NAE & NRC, 2002). Indeed, many technologies historically have been produced without the understanding of the underlying science concepts, such as the inclusion of a keystone in an arch. At this point, the STEM conceptions of the two teachers looked strikingly similar, but later analysis revealed that their understandings of STEM education were very different.

5.3.2 Knowledge for Teaching STEM

Wendy's and Charles' knowledge for teaching STEM is compared in this section according to the four knowledge domains: knowledge of instructional strategies, students, curriculum, and assessment. We infer their conceptions of the teacher's role in STEM education from their interview accounts.

5.3.2.1 Instructional Strategies for Teaching STEM

STEM education emphasizes collaboration as individual students may have different interests, expertise, and experiences. Charles regarded collaborative learning as important in STEM classrooms. Likewise, Wendy explained the importance in a more elaborated manner using the Maker program in her school as an example. She said:

Our Maker program involves students developing different products in groups [that met the social needs]. They have to collaborate since we believe every student has his/her talents. Everyone can contribute to the project.

While describing the collaborative opportunities offered by her school-based STEM program, Wendy actually wished to help her socially deprived students with low self-confidence to realize their potential in making their own products that should involve scientific and/or design principles. She stressed the importance of student-centeredness in her STEM curriculum by stating:

It must be student-centered. I won't spoon-feed the students; neither did I prepare for them the ingredients and ask them to cook with me [which was an analogy]. My goal [of STEM instruction] is to broaden their learning as much as possible. I have to be hands-off.

Wendy was alluding to the undesirability of teacher as a knowledge transmitter using a traditional, teacher-directed *cookbook* approach. Her students could decide on the products they wished to develop and change their mind within a time frame. Wendy did not wish to set a boundary with prescribed ingredients and cooking methods for students to carry out their projects. She also mentioned that there were no right or wrong approaches and solutions to the problems students were experiencing. Her hands-off, open approach provided a lot of space for students to nurture their creativity and exploratory mind.

Wendy brought this instructional approach to her disciplinary teaching by stating:

I have provided more space for student discussion in my biology lessons. I purposely let them brainstorm and argue based on the "no right or wrong" principle. For example, the biology topic of ecology covers a lot of news media where students can find information on their own. I can ask my students to find the reasons why many fish are found dead on a beach. They can propose whatever they like, and their ideas are then evaluated by the others. My students actually like this way of learning.

Wendy's experience with the junior secondary STEM curriculum transformed her teaching in senior biology classes to a certain extent. Despite acknowledging the constraints of the senior secondary curriculum (e.g., the tight teaching schedule and high-stakes examinations), she aimed to be a learning facilitator and support students to explore the more advanced scientific principles.

Charles stressed a similar role for STEM teachers, as learning facilitators who were more lenient and open-minded. He stated:

In STEM education projects, the students should lead the projects and the teacher can be their consultant. For example, the teacher can prompt the students to identify some problems [within their projects] and/or the possibility to modify [the designs]. ... Otherwise, the meaning of project-based or design-based learning will be lost.

What made Charles as a preservice teacher different from Wendy was his persistence in guiding students to link their STEM practices with scientific principles. From the start of the interview, Charles mentioned at least four times that such an approach is necessary to make STEM instruction effective. This was illustrated when he discussed the STEM lesson design:

You need to guide them to think about the *relationship between the theories and their product designs*. For example, you can require them to present their design principles and question about the [scientific] theories, no matter if it is simple or complex, that can support their argument. This will make it easier to *focus their attention back* to the linkage between the two.

An important feature of Charles' statements on the link between theories and students is the close reference between his understanding of STEM and STEM pedagogies. With a strong emphasis on his role as a science teacher who should primarily focus on teaching science (discussed in the previous section), Charles consistently put himself as a learning facilitator within a boundary; that is, his lessons should aim to scaffold students' understanding of scientific concepts at the end. His instructional approaches, when compared with Wendy's, were less open and had a more definite sense of what concepts are right.

In summary, both teachers identified some features of STEM teaching that included collaborative learning, student-centered classrooms, problem-solving, integrating design-and-make, and active inquiry into the scientific principles. The major differences between the experienced and novice teachers were how they interpreted the meaning of learning facilitators and the richness in their elaborated answers.

5.3.2.2 Knowledge of STEM Curriculum

Charles' perceived role of STEM teachers not only had an impact on his knowledge of instructional strategies but also affected his interpretation of the goals of the STEM curriculum. His perceived learning objectives were closely bound with disciplinary learning.

I *guess* in the attitudes domain [of the learning objectives] STEM lessons can help students to realize that they are *acquiring subject matter knowledge* that would be useful to solve real-life problems. Second, their *understanding of the theories* can be more solid. … When they make the products, they will understand how different actions such as changing a parameter would bring different results. … Finally, if science is not the emphasis, STEM activities will let students recognize how to use the technology.

Charles regarded understanding and application of scientific principles as of paramount importance in teaching STEM. STEM was more like a tool or a context that would be useful for science teaching. Although STEM involves technology, an understanding of the technological applications still dominates in his conceptions. He defines technology as a *product or a noun, and not a process or verb*. More importantly, Charles started his explanations by saying "I guess," which demonstrated how a lack of STEM teaching experience restricted his understanding of the new objectives of the reformed curriculum.

Wendy had the experience to develop the Maker program in her school. Although she prioritized science learning as did Charles, Wendy clearly identified learning generic skills as crucial in STEM education. She envisaged that to survive, students in the twenty-first century need to acquire certain skills and capabilities that include self-learning so that they can continue to learn outside the classroom, the capabilities to cope with adversity even if they fail the public examinations, and creativity to innovate so that humans will not be replaced by artificial intelligence in the future.

According to Wendy's experience in planning the STEM curriculum, she was different from Charles in that she could put forward a blueprint of the curriculum by considering her students' prior knowledge and skills. For example, she pointed out that her Grade 7 STEM lessons should focus on the skills and attitudes necessary for making the products, that Grade 8 should be a reinforcement stage on what students had experienced previously, and that Grade 9 needs to aim at nurturing entrepreneurship.

In the interview, Wendy noted the remarkable curricular differences between STEM and science. She used a "two world" analogy when comparing them:

> The biology curriculum and school-based Maker curriculum are *really different*. The biology curriculum was proposed by the Education Bureau, and the learning in the Maker program is decided by my students and me. I define the framework [e.g., project- and design-based]. Then my students decide on what they want to learn from the projects. I try to accommodate their requests. Therefore, the two [curricula] are *two worlds*.

In summary, Wendy's engagement in real-world STEM education had shaped her understanding of the STEM curriculum. STEM was not simply supporting the learning of science as opined by Charles. STEM, by its nature, provides more flexibility and room for students to take charge of their learning, that is, what and how they learn. Not only knowledge but also practices and habits of mind are the learning objectives of STEM. The school-based STEM curriculum should suit the specific needs of particular students. The drastically different conceptions of STEM curriculum held by Wendy and Charles are further evident from their knowledge of students in the next section.

5.3.2.3 Knowledge of Students

Wendy's practices in STEM education were grounded in her understanding of students and then further developed from her practical knowledge gained from running the Maker program. Most of her students came from socially deprived families and/or were new immigrants living in the same district. Their learning motivation was quite low. As a result, only a small proportion of students could enter university after Grade 12. Many would go into vocational training programs or even enter work without a clear career plan. Therefore, the Maker program aimed to improve their generic skills and attitudes. Wendy stated:

> *My aspiration* is to help students develop self-learning and cope with adverse conditions. These soft skills would benefit them in the long term. I really wish to use experiential learning, or guide the students in hands-on activities, so that they can find it appealing to learn.

Wendy's precise understanding of the weaknesses of her students had shaped her conception of how the STEM program should look in the school. By focusing on generic skills and less on scientific concepts and process skills, the weak students would find the environment conducive and comfortable to learning what they wished to learn. As a result, some students exceeded Wendy's expectations on their motivation to learn. She cited two examples:

> A group of students did not let me know that they had made an appointment with the external Maker laboratory so that they could continuously work on their product. Another example was the students read all the online information about wheelchair designs I had sent to them. They took the initiative reporting to me! Actually, I had never read the information before! [Wendy laughed]

When talking about her students, Wendy always used an affirmative tone. She was positive toward the students' successes, frustrations, and failures in the process of learning. Even when the students said it was impossible to use very limited resources to make the products (similar to most of the schools having STEM programs), she presented this as one of the task's design challenges. Students are expected to find solutions instead of making complaints to the teacher.

On the contrary, Charles provided a very limited discussion on students' learning in the interview. The only occasion was to express his concerns about the problems brought by a different mode of learning, that is, the impacts of having a student-centered STEM classroom. Charles said:

> Although many people encourage student-centered learning, our students are generally *knowledge recipients* due to the tight teaching schedule. They do not have many opportunities to develop problem-solving, critical thinking, or creativity. The students might be confused if they suddenly have to carry out design projects [STEM projects]. Therefore, the feasibility [of STEM teaching] is dependent on the capabilities and interests of the students.

Although putting himself as a learning facilitator (within a boundary), Charles's pessimistic view of making the classroom more student-centered was actually contradictory to what he had said. In his opinion, students would find it difficult or disinterested in taking active roles in learning if they wish to stay as knowledge recipients. As a result, teachers should regard themselves as knowledge dispensers instead of being facilitators of learning. Charles' perspective suggested that students' dispositions to learn would remain static and indifferent toward any changes in curriculum and instruction. Once again, this reveals Charles's narrow purposes for STEM education. Rather than the practices and attitudes, STEM teaching was merely to support students in learning scientific concepts for the sake of better academic performance. He stated:

> STEM education would be meaningful if it can help students develop understanding of scientific concepts, or increase their learning motivation in science, so that they would succeed in examinations.

Charles' predominant focus on learning scientific conceptions as the STEM curriculum goal was made clear.

5.3.2.4 Knowledge of Assessment

In theory, teachers' knowledge of assessment has a close relationship with the other knowledge bases. Their knowledge of curriculum, students, and instruction should interact and be consistent with their knowledge of assessment. Interestingly, both teachers shared similar views on STEM assessment although Charles and Wendy had considerable differences in their other knowledge bases.

Charles identified that formative assessment should be adopted in STEM teaching. In his opinion, the assessment needs to evaluate students' skills for their STEM projects. A paper-and-pen assessment would be inappropriate as the whole evaluation can have only one or two questions on solving problems. Charles's focus on assessing students' practices in STEM was contradictory to his expressed emphasis on content knowledge. This inconsistency may be accounted by Charles' limited experience in teaching STEM. He also did not suggest any specific examples of assessment for STEM.

Wendy had similar ideas as Charles to opt for formative assessment that focuses on the learning process and less on the final product. She provided more elaborations on what and how to assess in her Maker program. In line with the learning objectives, she assessed students' creativity, self-learning ability, capability to combat adversity, development of entrepreneurship, and being empathetic to those in need. A variety of assessment tools were employed. First, students kept portfolios that comprised all their designs, prototypes, and reflection reports. Teachers would review these portfolios at least twice a year. Second, a questionnaire was completed by the students three times per year. The questionnaire items were developed by taking items or tasks from currently available assessment tools (e.g., those on creativity), and some questions were specific for the school-based program. Third, teachers observed student learning progress regularly from the weekly lessons that were part of the school's formal timetable. Finally, students had to give group presentations after the completion of the project. Wendy regarded the first three assessment methods to be stress-free as students' performance in STEM would not be counted in their academic reports. The final presentation could be more stressful, but it remained as a platform for the students to demonstrate their work. In summary, Wendy's knowledge of STEM assessment was richer and more consistent than Charles from nurturing the students to evaluating their learning, and from providing less stressful, low-stakes assessment to increasing students' motivation to learn.

5.4 Discussion

This chapter examines the conceptions of a preservice teacher and an in-service teacher of STEM and their practical knowledge for STEM teaching. It clearly reveals how their perceived roles of STEM teachers and their experiences—in the midst of implementing a curriculum reform initiative—shaped the development of their knowledge for teaching STEM. Teachers' strong science background undoubtedly affected their conceptions about STEM, for which science always guides the development o f technology and engineering.

The findings show that preservice science teachers, like Charles who had limited exposure to STEM teaching except from a course, could still point out some domains in developing STEM literacy (Zollman, 2012), for example, the cognitive domain (e.g., knowledge and processes). Although Wendy seemed to stress more on the understanding of scientific concepts at the beginning of the interview, her latter account revealed that psychomotor goals (e.g., collaboration and communication) and affective objectives (e.g., students' persistence and attitudes toward difficult situations) were actually the ultimate objectives of the STEM programs. On the contrary, Charles consistently saw the goal of learning scientific concepts through STEM education and less on the other domains of STEM literacy. In other words, Charles appeared to hold an instrumental view about STEM education in which the other disciplines provide a context for learning of canonical science ideas.

Similar to previous studies on the development of teacher professional knowledge (e.g., Verloop, van Driel, & Meijer, 2001), teachers' experience in disciplinary teaching and STEM education is inevitably an influential factor affecting their conceptions and understanding of STEM. Being a biology teacher, Wendy tended to recognize living systems as interdisciplinary science. Her experience to teach junior grades might have predisposed her to emphasize processes and practices and de-emphasized the content outcomes. As a result, Wendy repeatedly stressed the importance of looking after the long-term needs of her students (e.g., my aspiration is to help students). She was able to illustrate the learning progress of her students with two specific examples (e.g., students approached the Maker laboratory on their own). Her understanding of students, in return, had made the school-based STEM program a stress-free and another world from the formal curricula as set out by the Education Bureau (i.e., the two-world analogy). Wendy's student-centered approach was featured by high flexibility and openness, which not only allowed her junior secondary students to decide on what and how to learn but also led to a revamp of her strategies in teaching senior secondary classes—even though it had a high-stakes assessment. Wendy clearly demonstrated that her knowledge was more specific to her students, more contextualized (e.g., the Maker program involving all junior secondary students is a unique case in Hong Kong), and more detailed (e.g., what and how the learning goals could be achieved).

It is worth pointing out that it is not our intention to identify Charles as an insufficient teacher since he was still at an early stage of exploration of what STEM teaching actually means. As a physics (and physics education) student, Charles had strong deductive strength that made him believe an individual could not innovate in technologies and engineering if s/he could not provide appropriate reasons and principles. His continuous emphasis on having STEM as a platform to teach scientific concepts (and less on the other aspects of science learning), when compared with Wendy's focuses of STEM education, was apparent. With his use of words (i.e., I guess) and a wobbling perceived the role of STEM teachers (i.e., from learning facilitator to knowledge dispenser and vice versa) in discussing his knowledge bases, Charles signified that the sources and nature of professional development should be different between preservice and in-service STEM teachers.

Charles relied on his undergraduate program to develop his understanding of STEM education including its purposes, effective pedagogies, and assessment strategies. Since novice science teachers with discipline-specific background may lack the necessary understanding of and learning experiences in some areas, particularly engineering designs, it is essential to equip them with the interdisciplinary content knowledge (Stohlmann et al., 2012). They need to learn how integrated STEM teaching would be possible from their university courses (Kelly & Knowles, 2016). Opportunities to talk and work with real engineers and technologists and related entrepreneurs would be useful. At the same time, these teachers should be given opportunities to practice what they have learned in classrooms, for instance, holding microteaching sessions in their courses and STEM workshops in their teaching practicum. In other words, the features of high-quality STEM education for school students (Kennedy & Odell, 2014) should also be found in the professional development of preservice teachers.

The learning needs of experienced STEM teachers can be different from the novices. Wendy was a more experienced STEM teacher with some basic training. However, she was still quite new to the STEM curriculum as the STEM movement had only been in place for 2 years at the time of the present study. She sometimes had questions and worries that she really wished to seek support from more knowledgeable others. Without much guidance, Wendy needed to explore on her own. Other than learning from various online courses, she sought help from professional consultants and connected herself with the broader STEM communities, in particular with those schools that had school-based STEM curricula. Through building networks and partnerships, the teacher learning communities became her major source of learning and support. Based on Wendy's experiences, supporting in-service teachers in establishing STEM communities and becoming self-directed learners could be a direction for future teacher professional development.

5.5 Future Work

To conclude, this study examined two science teachers' conceptions of STEM and their practical knowledge for teaching STEM. Our studies revealed that even though both teachers appeared to hold similar STEM conceptions, they indeed held very different nature and content of practical knowledge for STEM teaching. Further analysis of more STEM teachers, especially those in-service teachers with different disciplinary backgrounds, will be necessary for identifying the distinctive conceptions and knowledge of STEM teachers.

Another direction of research is to explore how the professional knowledge of preservice teachers can be improved for integrated STEM education. For example, what types of courses and course content would be necessary to strengthen their scientific, technological, engineering, and mathematical literacies as a whole? What learning experiences can help preservice teachers better bridge their understanding with practices? How can they become curricular leaders in STEM education in the future? Answers to these questions are imperative in raising the competency of teachers to teach STEM and, hence, to realize the reform goals of supporting students in developing STEM literacy.

Acknowledgements The authors wish to thank the teachers who participated in the study, Teaching Development Grants of The University of Hong Kong for the funding, and the reviewers' valuable feedback.

References

Berlin, D. F., & White, A. L. (1995). Connecting school science and mathematics. In A. F. Coxford & P. House (Eds.), *Connecting mathematics across the curriculum* (pp. 22–33). Reston, VA: National Council of Teachers of Mathematics.

Blackley, S., & Howell, J. (2015). A STEM narrative: 15 years in the making. *The Australian Journal of Teacher Education, 40*(7). https://doi.org/10.14221/ajte.2015v40n7.8.

Bybee, R. W. (2010). Advancing STEM education: A 2020 vision. *Technology and Engineering Teacher, 70*(1), 30–35.

Denzin, N. K. (1989). *The research act: A theoretical introduction to sociological methods.* Englewood Cliffs, NJ: Prentice Hall.

Dickerson, D. L., Cantu, D. V., Hathcock, S. J., McConnell, W. J., & Levin, D. R. (2016). Instrumental STEM (iSTEM): An integrated STEM instructional model. In L. Annetta & J. Minogue (Eds.), *Connecting science and engineering education practices in meaningful ways* (pp. 139–168). Cham, Switzerland: Springer.

Education Bureau. (2016). *Report on Promotion of STEM Education: Unleashing the potential of innovation.* Hong Kong: Author.

English, L. D. (2016). STEM education K–12: Perspectives on integration. *International Journal of STEM Education, 3*(3). https://doi.org/10.1186/s40594-016-0036-1.

Gess-Newsome, J. (2015). A model of teacher professional knowledge and skill including PCK: Results of the thinking from the PCK summit. In A. Berry, P. J. Friedrichsen, & J. Loughran (Eds.), *Re-examining pedagogical content knowledge in science education* (pp. 28–42). New York, NY: Routledge.

HKSAR Government. (2015). *The 2015 policy address.* Hong Kong: Author.

Honey, M., Pearson, G., & Schweingruber, H. (2014). *STEM integration in K–12 education: Status, prospects, and an agenda for research.* Washington, DC: National Academies Press.

Hurley, M. M. (2001). Reviewing integrated science and mathematics: The search for evidence and definitions from new perspectives. *School Science and Mathematics, 101*(5), 259–268.

Keefe, B. (2010). *The perception of STEM: Analysis, issues, and future directions.* Burbank, CA: Entertainment & Media Communication Institute, Division of Entertainment Industries Council.

Kelly, T. R., & Knowles, J. G. (2016). A conceptual framework for integrated STEM education. *International Journal of STEM Education, 3*(11). https://doi.org/10.1186/s40594-016-0046-z.

Kennedy, T. J., & Odell, M. R. L. (2014). Engaging students in STEM education. *Science Education International, 25*(3), 246–258.

Merriam, S. B. (2015). *Qualitative research: A guide to design and implementation* (4th ed.). San Francisco, CA: Jossey-Bass.

Morrison, J. S. (2006). *Attributes of STEM education: The student, the academy, the classroom.* Cleveland Heights, OH: TIES.

National Academy of Engineering & National Research Council. (2002). *Technically speaking: Why all Americans need to know more about technology* (G. Pearson & A. T. Young, Eds.). Washington, DC: National Academies Press. https://doi.org/10.17226/10250.

National Research Council. (2013). *Next generation science standards: For states, by states.* Washington, DC: National Academies Press. https://doi.org/10.17226/18290.

Patton, M. Q. (2002). *Qualitative research and evaluation methods* (3rd ed.). Thousand Oaks, CA: Sage.

Sanders, M. (2009). STEM, STEM education, STEMmania. *The Technology Teacher, December/January*, 20–26.

Stohlmann, M., Moore, T. J., & Roehrig, G. H. (2012). Considerations for teaching integrated STEM education. *Journal of Pre-College Engineering Education Research, 2*(1). https://doi.org/10.5703/1288284314653.

Verloop, N., van Driel, J. H., & Meijer, P. (2001). Teacher knowledge and the knowledge base of teaching. *International Journal of Educational Research, 35*(5), 441–461.

Yore, L. D. (2011). Foundations of scientific, mathematical, and technological literacies—Common themes and theoretical frameworks. In L. D. Yore, E. Van der Flier-Keller, D. W. Blades, T. W. Pelton, & D. B. Zandvliet (Eds.), *Pacific CRYSTAL centre for science, mathematics, and technology literacy: Lessons learned* (pp. 23–44). Rotterdam, The Netherlands: Sense.

Zeidler, D. (2016). STEM education: A deficit framework for the twenty first century? A sociocultural socioscientific response. *Cultural Studies of Science Education, 11*(1), 11–26.

Zemelman, S., Daniels, H., & Hyde, A. (2005). *Best practice: New standards for teaching and learning in America's schools* (3rd ed.). Portsmouth, NH: Heinemann.

Zollman, A. (2012). Learning for STEM literacy: STEM literacy for learning. *School Science and Mathematics, 112*(1), 12–19.

Chapter 6
Teacher Professional Development for STEM Education: Adaptations for Students with Intellectual Disabilities

Winnie Wing Mui So, Jia Li and Qianwen He

6.1 Introduction

Science, technology, engineering, and mathematics (STEM) education has received a great deal of attention; it has been positioned in many countries' development strategies (e.g., China, Singapore, South Korea, UK, USA), which are equally important to all learners regardless of their identity or ability. Hence, it is imperative to develop strategies that overcome barriers to ensure all students can achieve STEM literacy. This research aims to provide professional development (PD) support to teachers in the special school context to equip them with strategies in STEM education to help their students with special educational needs (SEN) so that they will not be deprived of STEM education opportunities. The term *special educational needs* has a legal definition in Hong Kong; it refers to children who have learning problems or disabilities that make it harder for them to learn than most children of that age (Education Bureau, 2014). Under the prevailing government policy, children with severe or multiple disabilities attend special schools where they are provided with intensive support services. The following sections discuss recent trends in STEM education including the importance of STEM education to all learners, the need to engage students with an intellectual disability (ID) in STEM education, teachers' roles in supporting ID students in STEM education, and professional development needs o f special school teachers in STEM education.

W. W. M. So (✉) · Q. He
Department of Science and Environmental Studies, Centre for Education in Environmental Sustainability, Education University of Hong Kong, Hong Kong, People's Republic of China
e-mail: wiso@eduhk.hk

J. Li
College of Life Sciences, Capital Normal University, Beijing, People's Republic of China

© Springer Nature Singapore Pte Ltd. 2019
Y.-S. Hsu and Y.-F. Yeh (eds.), *Asia-Pacific STEM Teaching Practices*,
https://doi.org/10.1007/978-981-15-0768-7_6

83

6.2 Importance of STEM Education for All Learners

STEM is important as it has become part of everyday life, regardless of race or ethnicity, level of ability, language spoken, gender, neighborhood, or geographic location (U.S. Department of Education & Office of Innovation and Improvement, 2016). The Ministry of Education in China has declared that adding STEM education into primary school curricula will ensure that every Chinese student has STEM literacy and is able to adapt to the future development of society (GET China Insight, 2017). In Hong Kong, the Education Bureau advocates that school STEM education is critical in ensuring that all young people are equipped with the skills and knowledge they will need to succeed (Education Council, 2015). Singapore's Prime Minister has emphasized the importance of STEM education because the skills are crucial to Singapore for the next 50 years (Lee, 2015). The South Korean Ministry of Education, Science, and Technology (2011) launched STEM education as a main policy for reorganizing the national curriculum and formulated strategies in promoting STEM education in primary and secondary schools. Hence, it is apparent that STEM education is equally important to all learners, irrespective of their identity or ability.

6.3 The Need to Engage Students with ID in STEM Education

The call for STEM education for all makes it imperative to develop strategies for children to overcome barriers and ensure that all students benefit from a good science education and achieve science literacy (National Science Teachers Association, 2017). Researchers have also suggested that STEM education is valuable for enhancing the quality of daily life for students, especially for those with disabilities (Hwang & Taylor, 2016; Obi, 2014).

Special schools in many parts of the world, including the US (Wehmeyer, Lattin, & Agran, 2001) and Hong Kong (Curriculum Development Council, 2001), are implementing the concept of *One Curriculum for All*, which includes ID students. The Hong Kong school curriculum for General Studies, which integrates science education, technology education, and personal, social, and health education, has an adapted version for students with ID (Education Bureau & Hong Kong Institute of Education, 2013).

However, STEM education is mostly designed for mainstream students. There is a concern that ID students might feel frustrated and, therefore, develop negative attitudes toward STEM learning, thus reducing their ability to access and comprehend scientific information as they progress through school (Davis, 2014; Lee & Erdogan, 2007). Moreover, students with disabilities are often discouraged from taking mathematics and science courses in middle and high school. Even when students with disabilities are enrolled in these classes, they are often not fully included in the rigorous work required to be successful in the classes and beyond (Obi, 2014). This may possibly be due to the lack of sufficient STEM education in the early years of schooling.

As stated in the manual published by the American Association on Intellectual and Developmental Disabilities (Schalock et al., 2010, p. 3), intellectual disabilities are defined as "significant limitations both in intellectual functioning and in adaptive behavior as expressed in conceptual, social, and practical adaptive skills" originating before the age of 18. In Hong Kong, in comparison with their peers, the global development of students with ID is delayed; they have marked disabilities in cognitive functioning in some areas, for instance, abstract and logical thinking, memory, short-term memory span, easily distracted, language expression, and other cognitive abilities/traits (Education Bureau, 2014).

Individuals with disabilities, including ID, are often evaluated and defined by other people solely based on their limitations and disabilities rather than on their capabilities, strengths, and broad range of interests and abilities (Obi, 2014). Many people assume that a disability limits an individual's ability to be educated; however, these students have the ability to be productive members of the workforce. It is suggested that targeting middle school students and developing their interest may help them acquire sufficient knowledge and confidence in STEM during high school, which in turn may help to increase the number of individuals going into further STEM education and STEM-related careers (Obi, 2014).

6.4 Teachers' Role in Supporting Students with ID in STEM Education

The general belief in most countries is that the life functioning of students with ID will improve if they are provided with appropriate supports. Therefore, teachers in special schools need to be prepared to teach all students with ID in STEM areas and adjust their teaching strategies based on students' needs (Wakeman, Karvonen, & Ahumada, 2013). In order to provide appropriate supports, it is critical for teachers to identify the difficulties and needs of ID students in STEM learning. According to the International Classification of Functioning, Disability, and Health model (World Health Organization, 2001), ID students' participation and achievement in STEM education are closely related to three multidimensional factors.

First, STEM education consists of science topics and interdisciplinary relationships ranging in difficulty, abstractness, and complexity that involve intellectual functions such as reasoning, planning, solving problems, abstract thinking, comprehending complex ideas, and learning from concrete experiences. ID students could have varying challenges or impairments in one or more of these cognitive and/or intellectual functions that create barriers to learning content-loaded subjects like science involving technical vocabularies—which are hard for ID students to understand and remember the meanings of technical terms and scientific concepts (Lee & So, 2015). Therefore, curricular adaptations have to enable the teacher to bridge the gap between the science curriculum demands and the student's intellectual disability. Behzad and Prabha (2017) suggested that effective adaptations could be made by finding changes that suit the specific cognitive or intellectual challenge faced by a student in each step of learning. Therefore, teachers play an important role during lessons. Curricular adaptations can be applied by modifying each aspect of the curriculum including goals, instructional materials, teaching methods, and evaluation approaches in such a way that the changes overcome the obstacles in each step of learning—**List your steps here**. For example, for learning objectives, teachers could sequence the objectives from simple to difficult or from single to multiple. Adaptations such as providing pictures or a real model and using colors to show similar and different features could be incorporated in the receiving information step of the lesson. In the processing information step, strategies such as repetitions or step-by-step presentation of ideas in the concept after the task or concept analysis phase can be used to help ID students to remember what they have learned.

Second, effective education should consider adaptive behavior that includes skills related to language, reading, writing, social responsibility, following rules, etc. Adaptive behavior, or adaptive functioning, refers to the skills needed for a person to live independently. People with ID show limitations in some areas of adaptive behavior and can find it difficult to cope with the challenges of life. Certain skills are important for adaptive behavior, and they are summarized into three categories: the learning of concepts, social skills, and daily living skills (Hong Kong Down Syndrome Association, 2016).

Third, the performance of ID students in STEM education can be influenced by their health and environmental factors, including their supports (Obi, 2014). ID individuals with weaker intellectual or analytical capability need clear and specific guidelines and to learn at their own pace, they need more time and opportunities to learn, and they need encouragement to try so as to build confidence and self-esteem.

An ID label does not "communicate a student's complete learning profile" (Friend & Bursuck, 2015, p. 136). The supports needed by ID students can be considered "fluid, continuous, and changing, depending on the person's functional limitations and the supports available within the person's environment, [rather than being] fixed [or] dichotomized" (Schalock et al., 2010, p. 110). Thus, the main difficulty in engaging ID students in STEM education is to identify their unfixed support needs. The support needs for ID students should be considered as arising from multidimensional factors including their intellectual functions, adaptive behavior, and the learning environment. However, teachers should also keep in mind that ID students are similar to non-ID students rather than being different from them; therefore, it is essential to see their strengths (Epstein, 2000; Shaywitz, 2003). Teachers who utilize the strengths of ID students can help them overcome their learning difficulties (Tomlinson & Jarvis, 2006). Experienced effective teachers in special schools are capable of identifying and understanding the strengths and needs of ID students in most subject areas. However, when it comes to STEM education, most teachers in special schools are not prepared to implement innovative, adaptive education approaches because they are not familiar with STEM education and have no idea how to design effective STEM learning activities for their students.

6.5 Professional Needs of Special School Teachers in STEM Education

DeJarnette (2012) encouraged exposure of elementary school students to STEM knowledge as foundational for their early education. Ironically, many elementary teachers have limited background knowledge, confidence, and efficacy for teaching STEM, which may hinder their students' STEM learning. El-Deghaidy and Mansour (2015) indicated all of the teachers in their study expressed concern that they felt underprepared to use STEM applications in their classroom. Teachers in special schools were no exception.

The association between teacher preparation to teach STEM and student achievement in STEM has raised the concern of educators and researchers about providing teachers' PD programs. PD is considered a key component in helping teachers develop knowledge, skills, and attitudes toward interdisciplinary teaching through the transformation process. Altan and Ercan (2014) found that teacher PD programs need to provide learning opportunities for teachers themselves in order to deepen their conceptual understanding, help them engage in scientific and engineering practices, and develop their appreciation of STEM to be essential in a community of knowledge builders (National Research Council, 2012). It seems realistic for special school teachers to adapt the implementation of STEM education for mainstream practices to the special school setting, which is similar to the approach used for the adapted General Studies curriculum learning content and activities, to meet the needs of ID students in Hong Kong.

6.6 Methods

With an aim to support special school teachers to overcome barriers to ensure students with ID can achieve STEM literacy, the methodology of this study included a specially designed teacher PD program. Teachers' views in written responses about their perception of STEM education with ID students and their opinions about adaptation of STEM activities for ID students were collected before and after the PD program.

6.6.1 Teacher Professional Development Program Design

The PD program was designed to equip special school teachers with basic understanding of integrated STEM education, activities used in the mainstream schools, and opportunities to adapt such instruction to the needs of ID students. The conceptual framework involved learning theories and pedagogies that lead to achieving key learning outcomes for integrated STEM education and an inquiry learning process that caters for ID students' learning needs.

Kelley and Knowles (2016) suggested that it is essential to consider science inquiry and engineering design as basic with technological literacy and mathematical thinking as auxiliary in integrated STEM activities. However, since the existing local school curriculum and subjects focus on the areas of science education and technology education, it is more natural for STEM education to center on science inquiry and the use of technology in making and designing components as the foundation for the STEM PD program. Furthermore, the PD program in this study was designed to focus on enriching special school teachers' understanding of STEM education and increasing their knowledge and capability in designing STEM activities for their students with mild ID. For instance, in the STEM activity "exploring the secret of light," the focus is on investigating holographic properties and the phenomenon of light reflection through watching holographic videos featuring 3-dimensional (3D) images of animals, the earth, and the planets, which are topics in the established school curriculum. Afterward, the participating teachers are required to perform the task of designing a do-it-yourself (DIY) hologram pyramid to facilitate watching a holographic video incorporating the consideration of screen size of the viewing device—cell phone or computer tablet.

Lee and So (2015) emphasized that introducing inquiry problems using stories is an effective approach, and Hwang and Taylor (2016) advocated integrating arts into STEM education to reduce difficulties for ID students. Hence, all activity designs were related to daily scenarios. Furthermore, model making is a constituent element of the PD program, such as constructing a model of a bionic butterfly in the inquiry about butterflies, building a LEGO® model of a shark with a moving tail in the inquiry about ocean pollution, and making a compass in the inquiry about direction.

Table 6.1 provides a summary of the activities designed for the integrated STEM PD program with consideration of science inquiry, technological literacy, engineering design, and mathematical thinking. The science inquiry opportunities provided for teachers included science phenomena, environmental conditions, morphology and behavior, renewable energy, and earthquake proofing. There were opportunities to enhance technological literacy with the use of tools for inquiry: coding/programming devices such as Micro:bit© (https://microbit.org/), App Inventor© (http://appinventor.mit.edu/explore/), and LEGO Education WeDo 2.0© (https://education.lego.com/en-us/elementary/intro/wedo2); measuring tools such as voltmeter and shaker; visual-related technology such as hologram pyramids and virtual reality (VR) videos; and 3D drawing devices such as Tinkercad™ (https://www.tinkercad.com/) and 3D printers. Engineering design and mathematical thinking as a way of making or creating products were also involved in the STEM activities, for instance, engineering design with smart and DIY devices and models that provided opportunities for mathematics thinking.

Table 6.1 Teacher professional development program with considerations of science inquiry, technological literacy, engineering design, and mathematical thinking

STEM content	Details
Science inquiry	Science phenomena including light reflection, sound, directions, and magnetic field
	Environmental condition including water temperature, ambient temperature, soil moisture, and light intensity
	Morphology and behavioral characteristics of butterflies
	Renewable energy conversion
	Earthquake proofing
Technological literacy	Coding/programming devices including Micro:bit, App Inventor, and LEGO WeDo 2.0
	Measuring tools including voltmeter and shaker
	Visual-related technology including hologram pyramids and VR videos
	3D drawing devices including Tinkercad and 3D printer
Engineering design	Electronic design of a piano app, e-compass, smart garden monitoring device, and block programming of a bionic shark
	Construction including a bionic shark, solar-powered boat, maglev car model, and quakeproof building model
	Tool design including DIY hologram pyramids and making of wind turbines
Mathematical thinking	Size, proportion, shape, weight, temperature, angular degree, gear ratio, and range of environmental conditions

6.6.2 Research Methodology

A pre/post survey was used to explore participating teachers' perceptions about the adaption of STEM education to ID students. How STEM education can be transferred from the PD program to special school teachers' teaching greatly depends on (a) teachers' perceptions of STEM education and whether they believe that there are advantages for ID students to engage in STEM and (b) teachers' readiness to adapt what was introduced in the PD program to future use in their class.

Twenty teachers participated in the PD program; all were informed of the research at the beginning of the program, and their informed consent was sought before distributing the questionnaire consisting of their background information and perceptions. These teachers' teaching experience ranged from a few years to more than 20 years. All participants held the teacher qualification in special education, either in bachelor, master, and/or postgraduate level university programs. They were subject area teachers in General Studies—a core subject at the primary level integrating science education, technology education, and personal, social, and health education—which is considered as the most relevant subject for implementing STEM education.

Data were collected from the participating teachers using a questionnaire involving written responses about their perceptions of STEM education before and after the 3-day PD program. The four questions asked were as follows:

- What are the objectives of STEM education?
- What are the advantages of STEM education for ID students?
- How can teachers help with ID students' STEM learning?
- How should STEM teaching be designed?

The differences and similarities in the teachers' perceptions were analyzed to document their initial perception and determine if teachers changed their perception of STEM after participating in the PD program. Toward the end of the PD program, the teachers were invited to identify which STEM activities could be modified and adapted for their mild ID students. For those more adaptable activities, the teachers were asked to describe and explain the modifications and adjustments required to meet the needs of their students with mild ID. They were also asked to identify and discuss the benefits of STEM learning to the ID students.

Teachers' responses were analyzed based on the thematic analysis approach (Braun & Clarke, 2006; Clarke & Braun, 2013). The thematic analysis revealed that the most frequently occurring items were objectives of STEM education for ID students, benefits of STEM education for ID students, how to support ID students with STEM learning, and STEM teaching design for ID students. The strategies proposed to adapt the STEM activities for ID students were coded and summarized to identify predominant, consistent, and noticeable themes. The coding was done by one researcher, and the other researchers verified the assertions by a random check of the codes.

6.7 Findings

6.7.1 Changes in Teachers' Perceptions of STEM Learning for ID Students

6.7.1.1 Objectives of STEM Education for ID Students

Teachers who participated in this PD program were asked to write their impression of STEM education before and after the workshop. About half (55%) of the teachers were able to list some objectives of STEM education prior to the workshop; however, all were able to identify the objectives after the workshop. This group of special school teachers' initial perception of the STEM education objectives for ID consisted of four aspects: knowledge understanding, problem-solving, creativity, and interest cultivation. After the workshop, more teachers indicated problem-solving, which increased from 25 to 70%; the other three aspects did not have noticeable changes (Fig. 6.1).

The workshop offered teachers a clearer understanding of STEM education objectives. For example, Teacher K, who did not have clear ideas about the objectives of STEM education before the workshop, provided clearer and detailed thoughts afterward.

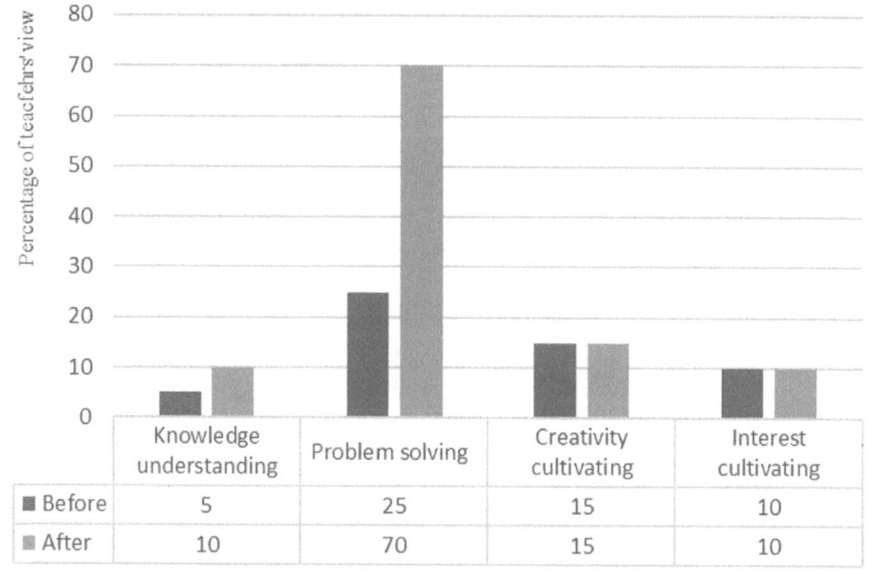

Fig. 6.1 Percentage of teachers' views about the objectives of STEM education for ID students based on responses from 20 teacher participants

> STEM education is interesting for students to learn; it is practical and is the general trend of education. (Teacher K, before workshop, without objectives)

> STEM education can cultivate students' ability to solve problems with creativity and innovations. Interdisciplinary thinking helps students seek different solutions by evaluating the effectiveness of the solutions and choosing the most appropriate way to achieve better outcomes. (Teacher K, after workshop, objective: problem-solving)

The workshop also partially changed and enriched some teachers' understanding of STEM education, especially about the purposes.

> STEM is closely related to our daily life, like the use of some technology products. STEM products are important and can help students understand the need to learn science and technology. (Teacher F, before workshop, without objectives)

> STEM education can help students gain more understanding of natural phenomena and people's needs in daily life and enable them to face the challenges in life. STEM education can improve the quality of students' lives. (Teacher F, after workshop, objectives: knowledge understanding, problem-solving)

Although some teachers had some initial understanding of STEM education, participating in the STEM PD helped them articulate the need for STEM education with ID students.

> STEM education can inspire students' interest in science inquiry and support students to learn and apply science principles and skills through different forms of activities. (Teacher P, before workshop, objectives: knowledge understanding, interest cultivating)

STEM education is an approach to assist students in interdisciplinary learning, which can give students more opportunities to participate in practices and to arouse their interest. (Teacher P, after workshop, objective: interest cultivating)

6.7.1.2   Benefits of STEM Education for ID Students

After the PD program, the teachers believed that the benefits of STEM education for ID students include problem-solving, followed by knowledge understanding, and higher order thinking. Other benefits include arousing interest, building learning ability, applying knowledge, cultivating creativity, and encouraging teamwork. Noticeable changes in applying knowledge (from 30 to 40%), problem-solving (from 45 to 65%), higher order thinking (from 10 to 35%), and cultivating interest (from 10 to 25%) were indicated after the program (Fig. 6.2).

Problem-solving was mentioned most frequently; however, some teachers did not realize this until their participation in the workshop. Also, teachers recognized the long-term impact of problem-solving ability on ID students' future life.

STEM education can help students integrate knowledge in different fields, which have life-long benefits. (Teacher C, before workshop, benefits: knowledge applying)

The advantage of STEM education is that it can assist students to master skills for problem-solving in daily life, know the relationship between learning and life, live independently, and have a high-quality life. (Teacher C, after workshop, benefits: problem-solving)

Initially, only 30% of the teachers had a general concept that STEM education could help students understand knowledge, while more teachers achieved clearer understanding of this after experiencing the PD on STEM in the workshop.

STEM education fosters students to understand more about the theories or principles and understand that there are many associations. (Teacher P, before workshop, benefits: knowledge understanding)

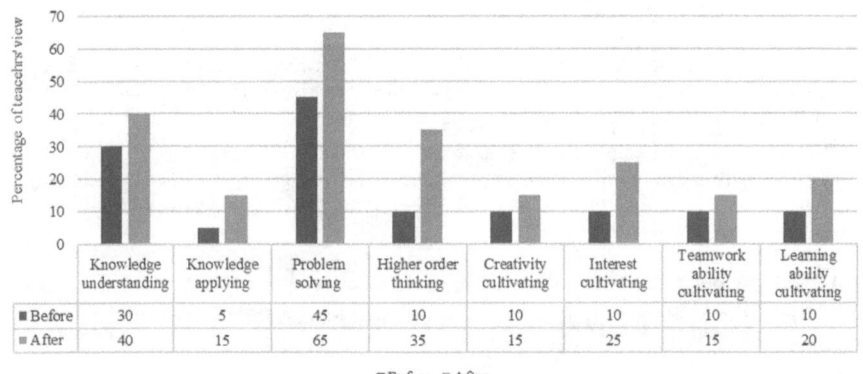

Fig. 6.2 Percentage of teachers' views about the benefits of STEM education for ID students based on responses from 20 teacher participants

ID students have fewer opportunities to participate in learning activities, but STEM education can remedy this situation. STEM education can provide situated learning activities that students can hardly experience in daily life, such as experiments, human body structure, space and other topics, or natural phenomena. (Teacher P, after workshop, benefits: knowledge understanding)

6.7.1.3 How to Support ID Students with STEM Learning

Before and after the workshop, the teachers were asked whether ID students have the ability to learn STEM. Surprisingly, all teachers believed the cognitive ability of most ID students enables them to engage in STEM learning. After the workshop, teachers had more confidence in saying "Definitely yes!" and gave more specific advice on the design of STEM learning for ID students. The teacher's role in teaching STEM for ID students is classified into the following four aspects of teacher support:

- Teachers should attend more STEM training programs to enrich their understanding of what STEM is and how STEM learning can be implemented.
- Teachers need to help students gain prior knowledge of STEM.
- Teachers should provide more practical opportunities for students.
- Teachers should observe the SEN of ID students and provide relevant guidance and assistance.

The number of teachers who suggested observing the SEN of ID students and providing guidance and assistance increased most obviously (from 25 to 50%), while the number of teachers who suggested attending more STEM training programs declined (Fig. 6.3). This implies that the PD program enabled these teachers to be more confident in teaching and designing STEM curricula.

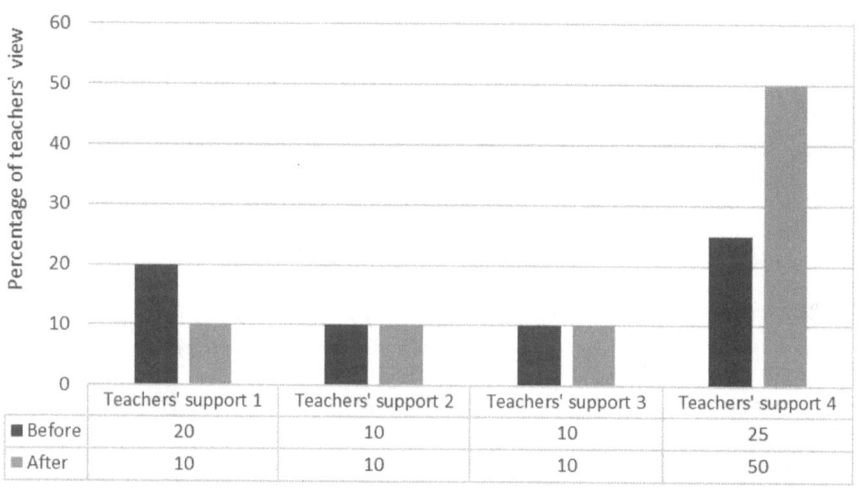

Fig. 6.3 Percentage of teachers' views about the teacher's role in teaching STEM for ID students based on responses from 20 teacher participants

It was found that most of the teachers initially did not have specific ideas to help ID students in STEM education; they only gave short and generalized responses. More detailed suggestions were given after the PD workshop.

> STEM learning needs to be adjusted for ID students and should be provided in different ways. Teachers should equip themselves with ideas in STEM learning and provide appropriate learning materials to encourage students to participate in STEM activities. (Teacher L, before workshop, teachers' support 1)

> We should choose the most suitable teaching materials according to students' ability. Teachers should consider and design the most appropriate activities, grouping the students properly, and design effective questions and summaries so that students can constantly reflect on and optimize their work during the STEM activities. Teachers also need to have a wealth of interdisciplinary knowledge to inspire students to think. (Teacher L, after workshop, teachers' support 1; teachers' support 4)

Teachers mostly gave suggestions based on their experiences in teaching General Studies before the workshop; however, they showed more targeted and practical recommendations for STEM education after the PD.

> Teachers should adjust the teaching content for ID students, for example, considering the situation where students' weak hand muscles affect their ability to do some work, like fastening and twisting. (Teacher S, before workshop, teachers' support 4)

> At the beginning of class, teachers should propose a problem close to students' daily life for them to think about ways to solve the problem. During the class, considering the limited time available, ID students can't finish the tasks only relying on their imagination, teachers should observe the students and give them necessary guidance and instruction. (Teacher S, after workshop, teachers' support 4)

Only a few teachers mentioned the support from teachers to help students gain prior knowledge of STEM. These responses were in a simple and short manner.

> Prepare students well with prior knowledge (Teacher E, before workshop, teachers' support 2)

> Assist students to build up prior understanding (Teacher R, after workshop, teachers' support 2)

Only a few teachers wrote about their opinion of teachers' support to provide more practical opportunities for students. Teachers N and H reiterated and emphasized their views after the PD program while Teacher A only mentioned this after the workshop.

> Provide ID students with more practical opportunities with STEM, enrich their experiences with STEM (Teacher N, before workshop, teachers' support 3)

> Give students more opportunities to try and let them master the inquiry cycle through more hands-on practices. (Teacher N, after workshop, teachers' support 3)

> More demonstrations, more practices, and let students create (Teacher H, before workshop, teachers' support 3)

> Provide semi-finished products, so that all students can have the opportunity to practice and finish the products by themselves. (Teacher A, after workshop, teachers' support 3)

6.7.1.4   STEM Teaching Design for ID Students

The analysis of teachers' responses (Fig. 6.4) found that their suggestions about STEM teaching design involve the following five teaching method (TM) components:

- The level of difficulty of the content should meet the level of the student's cognitive and hands-on ability.
- The STEM tasks should be broken down into smaller tasks for students to master step by step.
- The teaching content should relate to students' life experience.
- The teaching should be interesting.
- The preparation of appropriate aids and devices is necessary to cater to students' deficiencies and to protect them from danger in STEM practices.

The teachers' support for TM 1 increased from 30 to 40%, with 35% of teachers supporting TM 2, which was not realized by any of the teachers before this PD program. The teachers believed that it is easier for ID students to deal with small problems one by one, rather than one big complex problem. The percentage of teachers who supported TM 3 increased from 5 to 20%.

Due to the varied abilities of ID students, it is not difficult to comprehend that some teachers suggested the STEM learning content should meet the level of students'

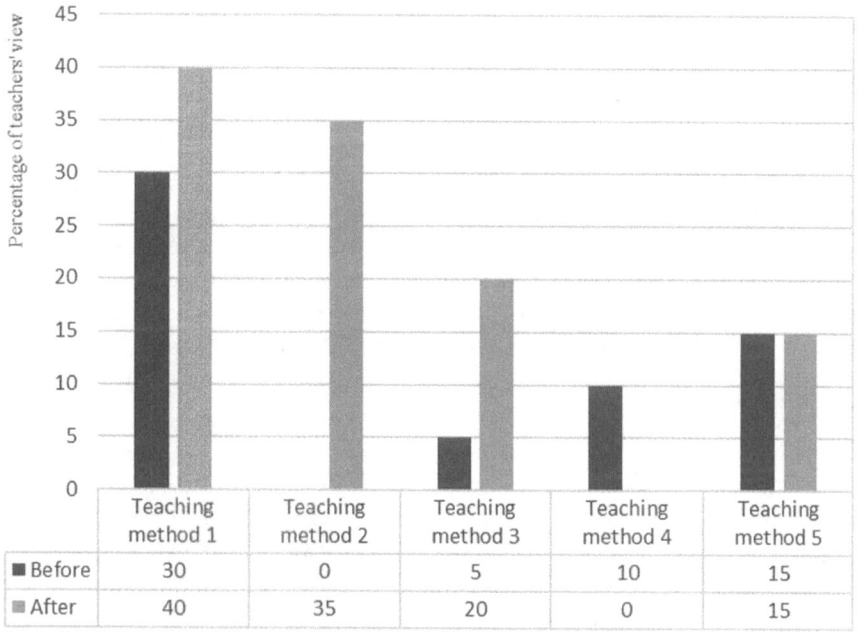

Fig. 6.4 Percentage of teachers' views about STEM teaching design for ID students based on responses from 20 teacher participants

cognitive and hands-on ability (TM 1). This opinion was mentioned most frequently both before and after the workshop.

> We can select suitable learning content and adjust it for ID students and develop suitable teaching materials for the ID students. (Teacher N, before workshop, TM 1)
>
> Teachers can design teaching contents according to students' abilities and explore the teaching effectiveness. (Teacher F, before workshop, TM 1)
>
> Teachers should choose the appropriate theme and design the STEM activities according to students' abilities and learning needs. (Teacher M, after workshop, TM 1)

None of the teachers realized that the STEM tasks should be broken into smaller tasks for students to master step by step (TM 2) before the PD program; however, 35% of the teachers proposed this suggestion after the workshop. This demonstrated that the PD program helped the teachers gain more knowledge about design of STEM activities.

> The STEM tasks can be broken down into small steps for students to work on the tasks step by step. (Teacher G, after workshop, TM 2)

The percentage of teachers suggesting the teaching content should be related to students' life experience (TM 3) increased from 5 to 20%. Teacher P was the only participant who mentioned this before the workshop, while four teachers stated this view after the workshop; their responses were similar.

> Teachers can start with phenomena and/or events that students are familiar with and incorporated with theories/principles. (Teacher P, before workshop, TM 3)
>
> Use situations that are close to students' daily life to facilitate them to immerse into the situations. (Teacher R, after workshop, TM 3)
>
> Connect the learning content with situations to make learning more related to daily life. (Teacher N, after workshop, TM 3)

Besides TM 3 that was focused on students' experiences, Teachers P and R initially suggested that teaching should be interesting (TM 4). However, after the workshop, they did not express this view about interesting; Teacher P changed to enhance the fun of STEM learning, and Teacher R changed to cater for students' interest. They started to pay more attention to students' learning. Moreover, teachers were writing about the view of preparing appropriate aids and devices to cater for the students' deficiencies (TM 5) before and after the workshop.

> [It is important] to equip students with different tools and learning strategies for the ID students to explore and inquire by themselves. (Teacher I, before workshop, TM 5)
>
> Use instrument that can show pictures and animations. (Teacher G, after workshop, TM 5)
>
> Teaching aids should be adjusted to make sure the students can use them safely. (Teacher T, after workshop, TM 5)

6.7.2   Teachers' Opinions About the Adaptation of STEM Activities for ID Students

The thematic analysis of the teachers' written responses to the following questions— *Do you think this STEM activity is adaptable for ID students? What are the benefits of STEM learning for ID students? How do you adapt the STEM activities for ID students?*—revealed that participants were able to provide suggestions for adaptation after their participation in the PD program.

6.7.2.1   Possibility of Adaptation of STEM Activities for I D Students

The percentage of teachers who indicated that STEM activities are adaptable was relatively high for most activities considered in the PD sessions: maglev car (100%), quake-proof building (100%), renewable energy (95%), inquiring about butterflies (95%), exploring the secret of light (90%), exploring marine pollution (85%), smart garden (85%), and making sound visible (75%). There were somewhat fewer teachers (45%) who considered that investigating directions by e-compass could be adapted for ID students.

The teachers' responses to the suitability of several technologies were somewhat varied. Most (90%) of the teachers considered that holographic images are suitable for ID students because they can support the learning of the characteristics of light, human organs, animal characteristics, etc.

> Holographic images can help ID students know about human organs, for them to feel like watching the real structure of human organs. Students can learn in groups with each student responsible for showing different organs. (Teacher G, adaptation: holographic images)

Most (85%) of the teachers believed that the mobile application Expedition and LEGO WeDo 2.0 are suitable for ID students since the teaching content could be designed according to the level o f he student's ability and knowledge.

> Students with high ability can try to build blocks, and other students can observe the building process. Expedition can guide students to learn relevant knowledge. (Teacher B, adaptation: Expedition and LEGO WeDo 2.0)

About 60% of the teachers considered that VR and 3D printing are useful for ID students while participating in STEM activities. In particular, VR technology would give ID students the opportunity to observe aspects of the world that they are unlikely to experience in daily life, such as the underwater world. Yet, 3D printing is slightly challenging for some ID students with low ability.

> VR can be used to experience different environments and to learn more about the characteristics of animals or other objects. (Teacher A, adaptation: VR and 3D printing)
>
> VR can be used for students to observe the marine organism and to tell the characteristics of the underwater world, and this is suitable for ID students in different levels. Students with better hand muscles can make some VR pictures and videos and share them with their classmates. 3D printing is suitable for older students to operate, because it is demanding to

use computer software and draw 3D graphics. 3D printing should be broken into small steps for students to follow. (Teacher I, adaptation: VR and 3D printing)

Using VR to observe butterflies is more suitable for students with mild mental retardation, while 3D printing is tough for them. (Teacher G, adaptation: VR and 3D printing)

However, over 75% of the teachers expressed concerns about the use of Micro:bit and App Inventor with ID students. They believed these two technologies are too demanding and beyond the ability of ID students.

In the use of Micro:bit to make a compass, it is difficult for the ID students to make it by themselves because this is beyond their understanding and hand–muscle coordination ability. It would be better if the teacher could prepare the Micro:bit compass for students' use. (Teacher Q, adaptation: Micro:bit and App Inventor)

Coding with App Inventor is rather difficult as students do not have much knowledge about coding. (Teacher B, adaptation: Micro:bit and App Inventor)

6.7.2.2   Benefits of STEM Learning for ID Students

In describing how to modify the STEM activities for ID students, the teachers also explained why they could be adapted. This is the key factor affecting teachers' intention to make suggestions for adapting these activities to the daily teaching and learning of ID students.

First, the teachers considered that these STEM activities could be focused on knowledge-oriented learning goals, which will help ID students achieve understandings about science, music, and computer science. For instance, these activities involve learning about natural phenomena of light characteristics, directions, force, magnet characteristics, and magnetic force and topics related to the environment and animals such as animal characteristics, features of the Earth and other planets, the marine environment, conditions for plant growing, renewable energy, earthquakes and building structures.

Second, these STEM activities are useful for improving the ability of ID students in aspects such as making and designing, logical thinking, strengthening muscles, and video-shooting. For example, building a shark using LEGO blocks and using an e-compass could improve the strength and flexibility of hand muscles. Students' attitudes and values (e.g., sense of responsibility, love of nature, and energy-saving notions) could be cultivated through learning about renewable energy and plant-growing conditions. STEM activities (e.g., coding with Micro:bit, designing and testing of maglev cars and quake-proof buildings) could improve students' logical thinking, scientific inquiry, and problem-solving skills.

6.7.2.3   Ways to Adapt the STEM Activities for ID Students

The teachers provided suggestions for each STEM activity to enhance their ID students' learning. The teachers mainly focused on the aspects of more steps to lower

Table 6.2 Suggestions for Adaptation/modification of STEM Activities for ID Students

Suggestions	Details/examples
Breaking down of tasks into smaller steps	Construction of a bionic shark robot using LEGO WeDo 2.0 can be broken down into many small steps with visual aids Step-by-step guides for assembly tasks will help students master the procedures
Dividing tasks in accordance with students' ability	Students with higher ability are more capable of coding with App Inventor and Micro:bit with less assistance needed, while those with lower ability can work on other tasks such as data collection or coding a few commands
Establishing background knowledge	Background knowledge of solar power such as the source of energy should be taught before students design the solar-powered boat
Providing relevant tools	Larger projectors can be used for the hologram activity in exploring the secret of light If possible, provide larger sized LEGO blocks for students who have weaker hand muscles
Attending to safety precautions	Students should be careful when carrying out cutting tasks in the maglev car activity

the cognitive load, distribution of tasks to address the needs of specific students, provision of background knowledge and relevant tools, and awareness of safety precautions. Table 6.2 shows the suggestions with details and examples provided by the teachers to modify the STEM activities for their mild ID students.

First, more advice and assistance with details on designing the STEM activities are needed for ID students with different cognitive and hands-on ability. Regarding the concern of differences in students' ability, the teachers suggested a division of tasks between students with higher and lower level ability. Teachers could do some of the assembly work in advance of class time and provide extra assistance for students with lower ability.

Second, teachers agreed that relevant background knowledge related to STEM activities should be provided to students before the actual activity. For provision of relevant tools, some teachers suggested that providing specialized tools (e.g., larger projectors, large building blocks, and colored print materials) would assist in observation and capturing of images.

Last, a few teachers mentioned concerns to ensure students' safe use of cutting tools in some assembly tasks. Furthermore, their experiences with STEM activities during the PD program inspired the teachers to design STEM activities specifically for ID students. For example, a campus treasure hunt game was suggested by one teacher by adapting and modifying the exploring directions activity even though not many teachers believed that investigating directions could be adapted for ID students.

6.8 Conclusions

Based on the integration of the existing key learning areas of science education and technology education in the school curriculum, this STEM education PD program for special education teachers emphasized science inquiry while using technology for making and designing with mathematical thinking. The PD sessions provided the teachers with relevant knowledge about STEM education used in mainstream schools and encouraged them to make appropriate modifications and adaptations to create learning activities suitable for ID students.

The teachers' perceptions of STEM education were positive and were enriched by attending the PD program. The problem-solving they experienced to complete the tasks helped them have a stronger sense that STEM education could improve their students' problem-solving ability. The teachers realized the benefit of STEM learning in supporting students in different aspects of knowledge application, higher order thinking, and cultivating interest. The peer-supported discussion and thinking about how to help ID students with STEM learning and the teachers' observations of these students' SEN led these teachers to recognize the importance of guidance and assistance in support of student learning (Wakeman et al., 2013).

Based on the findings focusing on teachers' suggestions for adaptation, special school teachers should take into consideration two factors. First, STEM learning activities should be designed to be closely related to science inquiry (Behzad & Prabha, 2017; So, Zhan, Chow, & Leung, 2018). Second, although it is important for ID students to know more about contemporary technologies, teacher support with the provision of procedures of the essential/manageable steps is helpful for ID students to complete science inquiry and engineering design tasks (Behzad & Prabha, 2017). The insights and experience from this research are of value to teachers involved in developing STEM education for ID students.

Acknowledgements The work described in this paper was partially supported by a grant from the Research Grants Council of the Hong Kong Special Administrative Region, China (EdUHK 18613118).

References

Altan, E. B., & Ercan, S. (2014). STEM education program for science teachers: Perceptions and competencies. *Journal of Turkish Science Education, 11*(1), 3–23.

Behzad, M., & Prabha, H. (2017). Curriculum adaptations in science for students with intellectual disability. *International Journal of Advance Research and Innovative Ideas in Education, 3*(1), 65–68.

Braun, V., & Clarke, V. (2006). Using thematic analysis in psychology. *Qualitative Research in Psychology, 3,* 77–101.

Clarke, V., & Braun, V. (2013). Teaching thematic analysis: Overcoming challenges and developing strategies for effective learning. *The Psychologist, 26*(2), 120–123.

Curriculum Development Council. (2001). *Learning to learn: The way forward in curriculum development*. Hong Kong: Author. Retrieved from https://www.edb.gov.hk/en/curriculum-development/cs-curriculum-doc-report/wf-in-cur/index.html.

Davis, K. E. B. (2014). The need for STEM education in special education curriculum and instruction. In S. L. Green (Ed.), *S.T.E.M. Education: Strategies for teaching learners with special needs* (pp. 1–20). Hauppauge, NY: Nova Science.

DeJarnette, N. (2012). America's children: Providing early exposure to STEM (science, technology, engineering and math) initiatives. *Education, 133*(1), 77–84.

Education Bureau. (2014). *Operation guide on the whole school approach to integrated education* (3rd ed.). Hong Kong: Author. Retrieved from http://www.edb.gov.hk/attachment/en/edu-system/special/support/wsa/ie%20guide_en.pdf.

Education Bureau & Hong Kong Institute of Education. (2013). *Supplementary guide to the curriculum guide on general studies for students with intellectual disabilities (P.1 to S.3)*. Hong Kong: Author. Retrieved from http://www.edb.gov.hk/attachment/tc/curriculum-development/major-level-of-edu/special-educational-needs/basic-edu-curriculum/Adapted%20Curriculum_15Oct2013.pdf [in Chinese].

Education Council. (2015). *National STEM school education strategy 2016–2026: A comprehensive plan for science, technology, engineering and mathematics education in Australia*. Carlton South, Australia: Author. Retrieved from http://www.educationcouncil.edu.au/site/DefaultSite/filesystem/documents/National%20STEM%20School%20Education%20Strategy.pdf.

El-Deghaidy, H., & Mansour, N. (2015). Science teachers' perceptions of STEM education: Possibilities and challenges. *International Journal of Learning and Teaching, 1*(1), 51–54.

Epstein, M. (2000). The behavioral and emotional rating scale: A strength-based approach to assessment. *Assessment for Effective Intervention, 25*(3), 249–256.

Friend, M., & Bursuck, W. D. (2015). *Including students with special needs: A practical guide for classroom teachers*. Upper Saddle River, NJ: Pearson Education.

GET China Insights. (2017, May 3). The STEM education in China: There's a long way to go [Web log message]. Retrieved from https://medium.com/@EdtechChina/the-stem-education-in-china-theres-a-long-way-to-go-7e67a2c439f4.

Hong Kong Down Syndrome Association. (2016). Retrieved from http://www.hk-dsa.org.hk/resources/id/?lang=en.

Hwang, J., & Taylor, J. C. (2016). Stemming on STEM: A STEM education framework for students with disabilities. *Journal of Science Education for Students with Disabilities, 19*(1), 39–49.

Kelley, T. R., & Knowles, J. G. (2016). A conceptual framework for integrated STEM education. *International Journal of STEM Education, 3*(11), 1–11. https://doi.org/10.1186/s40594-016-0046-z.

Lee, M., & Erdogan, I. (2007). The effect of science-technology-society teaching on students' attitudes toward science and certain aspects of creativity. *International Journal of Science Education, 11*, 1315–1327.

Lee, P. (2015, May 8). Science, technology, engineering, math skills crucial to Singapore for next 50 years: PM Lee. *The Straits Times*. Retrieved from https://www.straitstimes.com/singapore/education/science-technology-engineering-math-skills-crucial-to-singapore-for-next-50.

Lee, T. T. H., & So, W. W. M. (2015). Inquiry learning in a special education setting: Managing the cognitive loads of intellectually disabled students. *European Journal of Special Needs Education, 30*(2), 156–172.

Ministry of Education, Science and Technology. (2011). *The second basic plan to foster and support the human resources in science and technology (2011–2015)*. Seoul, South Korea: Author.

National Research Council. (2012). *A framework for K-12 science education: Practices, crosscutting concepts, and core ideas*. Washington, DC: National Academies Press.

National Science Teachers Association. (2017). *NSTA position statement: Students with exceptionalities*. Arlington, VA: Author. Retrieved from http://static.nsta.org/pdfs/PositionStatement_Exceptionalities.pdf.

Obi, S. O. (2014). Working with learners with cognitive disabilities in STEM. In S. L. Green (Ed.), *S.T.E.M. education: Strategies for teaching learners with special needs* (pp. 37–48). Hauppauge, NY: Nova Science.

Schalock, R. L., Borthwick-Duffy, S. A., Bradley, V. J., Buntinx, W. H. E., Coulter, D. L., Craig, E. M., … Yeager, M. H. (2010). *Intellectual disability, definition, classification, and systems of supports* (11th ed.). Washington, DC: American Association on Intellectual and Developmental Disabilities.

Shaywitz, S. (2003). *Overcoming dyslexia: A new and complete science-based program for reading problems at any level*. New York, NY: A.A. Knopf/Random House.

So, W. W. M., Zhan, Y., Chow, S. C. F., & Leung, C. F. (2018). Analysis of STEM activities in primary students' science projects in an informal learning environment. *International Journal of Science and Mathematics Education, 16*(6), 1003–1023.

Tomlinson, C. A., & Jarvis, J. (2006). Teaching beyond the book. *Educational Leadership, 64*(1), 16–21.

U.S. Department of Education & Office of Innovation and Improvement. (2016). *STEM 2026: A vision for innovation in STEM education*. Washington, DC: Author.

Wakeman, S., Karvonen, M., & Ahumada, A. (2013). Changing instruction to increase achievement for students with moderate to severe intellectual disabilities. *Teaching Exceptional Children, 46*(2), 6–13.

Wehmeyer, M. L., Lattin, D., & Agran, M. (2001). Achieving access to the general curriculum for students with mental retardation: A curriculum decision-making model. *Education and Training in Mental Retardation and Developmental Disabilities, 36*(4), 327–342.

World Health Organization. (2001). *International classification of functioning, disability and health (ICF)*. Geneva, Switzerland: Author. Retrieved from http://www.who.int/classifications/icf/en/.

Chapter 7
Teaching Engineering-Focused STEM Curriculum: PCK Needed for Teachers

Szu-Chun Fan and Kuang-Chao Yu

7.1 Introduction

The cultivation of creativity, design thinking, problem-solving, critical thinking, communication and coordination, and teamwork skills through interdisciplinary science, technology, engineering, and mathematics (STEM) courses has recently become a crucial issue in international education reform. Long-term departmental and examination-oriented teaching patterns in the education system have hindered students from applying their knowledge to solving real-world problems, particularly in the fields of science, technology, and engineering. Education research reports from around the world have detailed these problems (Blackley & Howell, 2015; Business Roundtable, 2005; National Research Council [NRC], 2011; Sorenson, 2010; Thomasian, 2011). Numerous countries have thus endeavored to implement reform programs based on interdisciplinary STEM education with the objective of increasing student interest in STEM subjects, cultivating STEM literacy among students, and developing their ability to face future real-world problems to enhance national competence (Pitt, 2009; Thomasian, 2011). The educational philosophy of STEM centers on connecting acquired knowledge with real-world challenges through purposeful problem-based learning and design processes. By increasing the ability of students to discover, explore, and solve problems, STEM education has the potential to produce modern citizens equipped with problem-solving and critical thinking competencies (Bybee, 2013).

Engineering design is the key to integrating STEM curricula. Engineering is a design that must satisfy certain conditions, meet human needs, and solve problems,

S.-C. Fan (✉)
Department of Industrial Technology Education, National Kaohsiung Normal University, Kaohsiung, Taiwan
e-mail: scfan@nknu.edu.tw

K.-C. Yu
Department of Technology Application and Human Resource Development, National Taiwan Normal University, Kaohsiung, Taiwan

© Springer Nature Singapore Pte Ltd. 2019
Y.-S. Hsu and Y.-F. Yeh (eds.), *Asia-Pacific STEM Teaching Practices*,
https://doi.org/10.1007/978-981-15-0768-7_7

whereas technology is the process and the final result of such design. During the engineering design process, engineers must integrate relevant scientific knowledge, mathematical calculations, and technological applications in a procedure that begins with a design concept and ends with an actual product. Therefore, engineering design is regarded as the optimal approach to the design of integrated STEM curricula, combining knowledge of diverse subjects through *engineering*. Teaching models—such as project-based, problem-oriented, and situated learning (Morrison & Bartlett, 2009; Sorenson, 2010)—provide a learning experience that develops worldwide vision.

The conventional teacher education process, however, focuses on cultivating subject-specific competency, rarely considering the concept of integrating interdisciplinary curricula. Moreover, teachers usually instruct in one subject only and lack understanding of other subjects. Although teachers acknowledge the importance of STEM curricula, they do not necessarily know how to start incorporating STEM into their non-STEM curricula. Custer and Daugherty (2009) suggested that the large knowledge structure behind highly integrated STEM curricula poses severe challenges to the planning of teacher education programs and professional development. Asunda (2012) noted that, because STEM curricula cover a wide range of general knowledge, defining and planning teaching content and methods as well as implementing instructional evaluation may be difficult. Wells (2013) explicitly indicated that conventional teacher education courses are unable to provide teachers with sufficient subject expertise and confidence for delivering integrated interdisciplinary curricula. Therefore, assisting teachers with their professional knowledge and competence regarding STEM education is extremely crucial to the outcomes of STEM curriculum implementation (Love & Wells, 2018).

Teachers are the major factor affecting the success of educational reform and promotion of teaching philosophy (Fishman, Marx, Best, & Tal, 2003). For STEM curricula focusing on interdisciplinary integration, the planning of various aspects—including curriculum development, material design, teaching and assessment strategies, and even the performance of onsite teaching—is a key element requiring careful consideration. Teachers should thus learn the crucial points for planning engineering design courses, the design of instructional methods and materials, and the application of teaching strategies and assessment techniques. Moreover, they should be familiar with the pedagogical knowledge of each STEM subject if they are to change their departmentalized teaching model to the STEM teaching method centered on engineering design, which guides students regarding the integration and application of interdisciplinary knowledge. The objective of this chapter is to explore the pedagogical content knowledge (PCK) required for the implementation of engineering-focused STEM curricula; the findings serve as a reference for teacher education and professional development.

7.2 Theoretical Foundation and Modules of Engineering-Focused STEM Curricula

7.2.1 The Nature of STEM Curricula

Comprising the required knowledge, attitudes, and skills for solving real-life problems, STEM literacy is an indispensable capability in the twenty-first century (Thomasian, 2011). STEM literacy requires comprehending and interpreting knowledge of science, technology, engineering, and mathematics in the natural and artificial worlds, based on which constructive and rational care and reflections may be proposed (Bybee, 2013). In modern society, the development of most scientific innovations, engineering designs, and technology products relies on the integration and application of science, technology, and mathematics. From the perspective of subject content, science is cognition of the nature, knowledge, and concepts of science, with an emphasis on exploring the principles of nature, developing the ability to make objective decisions, and determining scientific issues or development from the perspectives of science and technology (NRC, 1996). Technology refers to the capability of using, managing, and assessing technology, including abilities of comprehension, application, critical thinking, problem-solving using technological knowledge, and even participation in major technology-related decision-making (International Technology & Engineering Educators Association [ITEEA], 2007). Mathematics refers to the ability to interpret and creatively apply mathematics for making evidence-based judgments as well as for analyzing or solving problems through mathematical thinking (American Association for the Advancement of Science, 1993). Finally, engineering indicates the comprehensive use of scientific, technological, and mathematical knowledge for efficiently solving problems or satisfying needs in the real world (Asunda, 2012). A few STEM curricula in recent years have been developed focusing on engineering design; for example, Project Lead The Way engineering courses (https://www.pltw.org/our-programs/pltw-engineering), Engineering byDesign™ Program (https://www.iteea.org/STEMCenter/EbD.aspx), and Engineering is Elementary curriculum (https://www.eie.org/eie-curriculum).

7.2.2 Engineering-Focused STEM Curriculum: The Engineering Design Process and Its Elements

Engineering design is the process by which engineers put their ideas into practice under the limitations of certain specifications and conditions. The fundamental concept of engineering design is to solve unstructured problems, meet restrictive conditions, create and develop various solutions, conduct analyses, make effective design decisions, perform objective assessments, identify the optimal design, and consider the design's potential consequences (Atman et al., 2007; Carr, Bennett,

& Strobel, 2012). Therefore, engineering-focused STEM curricula are driven by situated learning that engages students in the systematic problem-solving process through engineering design activities. Scientific inquiry is integrated into the activity process and supplemented by mathematical knowledge to complete the analysis and assessment required in the engineering design process (Kelley, 2010).

The *Next Generation Science Standards* (NRC, 2013) defined engineering design as three practices: (a) the definition and delimiting of an engineering problem—exerting utmost effort to clarify and fulfill the conditions, constraints, or relevant problems required for solving the problem; (b) the design of possible solutions—generating various solutions and then analyzing and evaluating the probability of success for each solution before selecting the optimal option; and (c) the optimization of the design solution—evaluating the significance of functions through systematic testing to optimize the final solution. Mentzer (2011) proposed six elements in the process of engineering design, namely, problem definition, solution development, analysis/modeling, experimentation, decision-making, and teamwork. ITEEA (2007), Budynas and Nisbett (2011), and Householder and Hailey (2012) concluded that the engineering design process in terms of technology education encompasses the following:

- need for identification and problem definition,
- ascertaining standards and restrictive conditions,
- research and idea generation,
- potential solution exploration,
- synthesis for method selecting,
- modeling and simulation,
- development of a detailed design for the program,
- testing and evaluation,
- specification review,
- design optimization, and
- communication with others to elaborate upon the production process and results.

Carr et al. (2012) summarized the consensus on the core concepts of *doing engineering* based on an analysis of K–12 STEM competence indicators and standards in the United States as follows:

- identification of criteria, constraints, and problems;
- evaluation, redesign, and modification of products or models;
- evaluation of solution effectiveness;
- design of a product or process to solve the problems;
- offering reasons for proposing the design and solution;
- production of models, prototypes, and sketches;
- design of products and systems;
- selection of appropriate materials, optimal solutions, or effective methods;
- explanation of the factors crucial to the solutions and design;
- establishment of plans, layouts, designs, solutions, and processes;
- creation of solutions, prototypes, and graphics;

- discussion of problems, designs, or solutions;
- conduct a briefing on the solutions and designs;
- definition of the problems;
- brainstorming to develop solutions, designs, and plans and identify design issues;
- construction of designs, prototypes, and models;
- application of criteria, constraints, and mathematical models;
- improvement of solutions or models; and
- proposal of elements such as flowcharts, system planning, solution design, blueprints, and production processes.

These perspectives reveal that, although researchers have proposed diverse procedures described at distinct levels of detail, an in-depth review of the elements and connotations indicates mutual core concepts (Lewis, 2005; Merrill, Custer, Daugherty, Westrick, & Zeng, 2009; NRC, 2009, 2013). The present chapter deemed that engineering-focused STEM curricula may achieve interdisciplinary applications through a process comprising the following four segments: (a) the inquiry stage—problem definition and analysis; (b) the development stage—solution planning and forecasting; (c) the practice stage—modeling and testing of final products; and (d) the reflection stage—review and optimization (Fig. 7.1). Teachers should have a clear understanding of the connotations of the four segments when designing an appropriate teaching process.

Fig. 7.1 Flowchart of engineering design

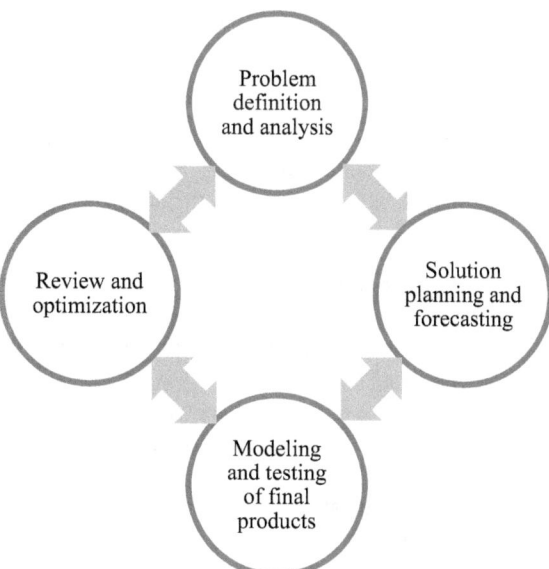

7.3 The PCK of STEM Subjects

7.3.1 Definition and Theoretical Foundation of PCK

PCK is the core of teacher knowledge and the basis of effective teaching. During the teaching process, teachers should determine how to organize, characterize, and transform particular topics according to the capabilities and interests of the learners to ensure the advancement of teaching and achievement of the learning goals (Shulman, 1986, 1987). Shulman divided the fundamental professional teaching competencies into the knowledge of general education and specific subject matter. General knowledge of education or pedagogical knowledge (PK) involves general understandings about teaching and assessment. Knowledge of subject matter comprises (a) content knowledge (CK)—the implications of subject expertise; (b) PCK—the conversion of subject expertise into teachable, learnable, and comprehensible knowledge; and (c) curricular knowledge—the integration and organization of the materials corresponding to relevant subject expertise according to the diverse levels of learners to develop specific course content and the teaching process. The PCK summit model proposed by Helmes and Stokes (2013) and Gess-Newsome and Carlson (2013) posited that teachers should have general, non-subject-specific professional knowledge (i.e., PK) as well as specific professional knowledge on a particular subject (i.e., CK). Thus, PCK is a combination of content knowledge and pedagogical knowledge. The objective of PCK is to determine how a particular topic, problem, or issue can be integrated, presented, and converted into teachable, learnable, and comprehensible knowledge to be acquired by learners with varying interests or abilities.

Widely accepted and discussed, PCK remains relatively abstract and is conceptualized differently for each subject. In the theoretical frameworks proposed by various researchers, PCK is generally classified into seven major dimensions, which serve as a framework for exploring the PCK needed for teaching STEM:

- Curriculum-oriented knowledge—knowledge related to the form and structure of the overall curriculum and teaching activity design.
- Knowledge of educational purposes and curricula—the relevant knowledge of the development of disciplinary content, organizational structure, and teaching objectives.
- Content knowledge—the conceptual and procedural knowledge of the discipline.
- Knowledge of teaching strategies—the strategies used in the teaching process to help students learn.
- Knowledge of educational context—knowledge with which teachers create a suitable teaching environment.
- Knowledge of learners—the teachers' prior knowledge of learners and their conceptions prior to commencement of a new unit.
- Knowledge of learning assessment—teachers using appropriate assessment tools for evaluating students' learning outcomes.

7.3.2 Characteristics of Conventional PCK in STEM

Overall, the uniqueness of PCK is related to the characteristics and presentation of subject expertise and each subject conventionally has its own PCK (Park & Oliver, 2008; van Dijk & Kattmann, 2007). The critical aspect is the differences caused by the knowledge of educational purposes and curricula, content knowledge, and curriculum-oriented knowledge. Such differences affect teaching strategies and the evaluation of learning outcomes. In addition, each subject requires different knowledge (e.g., planning and management of the laboratory and technology classroom) in teaching scenarios. For instance, science education focuses on guiding students to explore natural phenomena and principles. Conventional science education generally instructs in the outcomes of rigorous empirical research, and the teaching is based on lecture, demonstration, and guided inquiry. By contrast, learner-centered instruction has been prevalent in recent years and has orientations—such as process, activity-driven, discovery, project-based, inquiry-based, and guided inquiry based—for science education courses (Magnusson, Krajcik, & Borko, 1999).

Technology and engineering education focus on the cultivation of students' literacy in comprehending, selecting, and evaluating STEM knowledge in practices. The ITEEA (2003) proposed that technology teachers should comprehend the facts and skills taught in science and technology courses, understand learners' opinions regarding technology, be able to design and evaluate technology courses, adopt appropriate teaching strategies, design and manage an effective teaching environment and space for technology courses, and be responsible for their continual professional development. Thus, from the perspective of curriculum orientation, technology education often takes the form of practical skill development as well as orientation to designs, projects, problems, and interdisciplinary integration.

7.4 PCK for Teaching Engineering-Focused STEM Curricula

The integration of STEM content knowledge, which is applied in the practical and analytical aspects of engineering and technology classrooms, will depend on the ability of teachers to transform this knowledge into appropriate instruction for different grade levels (De Miranda, 2017). Engineering design incorporates various disciplines and is taught in an organized and logical multidisciplinary structure. Teachers should create a learning scenario that involves multidisciplinary knowledge and closely resembles a real-world situation, assisting students to create their own conceptual and procedural knowledge (Berry et al., 2004; Pinelli & Haynie, 2010; Toulmin & Groome, 2007). Due to variation in the nature of disciplines, science has a well-established epistemology leading to an established organization of knowledge. Technology, on the other hand, has no commonly agreed upon epistemology. Devising an effective approach to integrating various disciplines is a crucial challenge in the design of STEM curricula (Lantz Jr., 2009). STEM curricula can have diverse designs and forms depending on the particular course content, learners, and instructional objectives. To further explore this topic, this chapter details the PCK teachers should possess when implementing engineering-focused STEM curricula on the basis of the proposed four segments of the engineering design process.

7.4.1 Curriculum-Oriented Knowledge

Engineering-focused STEM curricula can generally be divided into content-based and context-based forms. Content-based curricula target content knowledge, instructing through practice activities. Context-based curricula do not adopt conventional departmental teaching forms and frameworks but adopt engineering design projects as their main focus for introducing relevant scientific, mathematical, and technological knowledge within the design process (i.e., from formulating a design concept to final production); this type of curriculum creates coherent intersections between disciplines (Herschbach, 2011; Kertil & Gurel, 2016). Prior to developing engineering-focused STEM curricula, teachers should familiarize themselves with the elements and implementation process of engineering design, thus enabling them to select the most appropriate curriculum type according to their teaching considerations.

7.4.2 Knowledge of Educational Purposes and Curricula

STEM places emphasis on giving students meaningful learning experiences through interdisciplinary integrated learning, inducing interest in learning. When delivering engineering-focused STEM curricula, teachers may convert concepts in the professional engineering field or contemporary social issues into project topics and problem scenarios. However, project-based activities require extensive and integrative knowledge frequently not yet possessed by students. Teachers should be familiar with the category and outline for the relevant disciplines during teaching. Moreover, teachers should comprehend the purposes of STEM curricula to establish explicit instructional objectives, thereby ensuring the relevance of each instructional activity.

7.4.3 Content Knowledge

Science, mathematics, technology, and other content knowledge play large roles in the engineering design process. Reimers, Farmer, and Klein-Gardner (2015) highlighted that teachers of engineering should clarify how engineering design offers a context for learning in science, mathematics, and other subjects. Teachers have the responsibility of ascertaining the characteristics and related content of each subject and should instruct students how to apply their relevant knowledge to the design process. During the inquiry stage, teachers should guide students to conclude the criteria and constraints of the problem (Fig. 7.1). Subsequently, students learn to break down a large problem into smaller, easier problems. This process usually requires the use of information technology for data collection and mathematics-related abstract thinking. During the stage in which solutions are devised, scientific and mathematical knowledge is needed for modeling, simulation, and predictive analysis. In addition, product production, testing, and modification during the practice stage require the integration and application of technological and scientific knowledge. The reflection stage involves scientific experiments and mathematical data analysis. However, most current K–12 STEM teachers are ill-equipped with multidisciplinary expertise. Interdisciplinary teacher communities or teams may be required to satisfy the need for this diverse CK.

7.4.4   Knowledge of Teaching Strategies

A STEM curriculum is an educational philosophy based on empiricism and constructivism; therefore, the instructional strategies employed in such curricula should center on learners so as to provide a meaningful inquiry and learning experiences as well as to emphasize knowledge construction and transformation. Aranda et al. (2018) noted that in an engineering design-based science curriculum, teachers should provide an interactive environment where students can fully discuss their ideas and receive rich responses to their questions and practices. More specifically, in the inquiry stage, teachers should use teaching strategies that involve clear task objectives, didactic teaching, problem posing and discussion, and role-playing. In the development stage, teaching strategies involving creative thinking, guided inquiry teaching, and cooperative learning should be used. In the practice stage, teachers should guide students through processes of inquiry experiments, practical learning, and technology-related problem-solving. Finally, in the reflection stage, mechanisms such as critical thinking, role-playing, and peer assessment should be employed. The particular strategies used should, however, depend on the actual teaching conditions of each teacher.

7.4.5   Knowledge of Educational Context

The learning environment used for the implementation of engineering-focused STEM curricula should provide students with experience in diverse inquiries as well as support and assistance during practice activities. Furthermore, hardware, conventional practices, and digital design may be integrated with production equipment to create a practice space resembling a maker space (Saorín et al., 2017), providing the opportunity for students to conduct self-directed learning. Knowledge of space management and equipment use is crucial for teachers. Teachers should create learning scenarios with a positive, open learning atmosphere that encourages exploration and takes on complex issues. Such an environment needs to grant students enough time to explore, test, and modify (Taylor, 2016).

7.4.6 Knowledge of Learners

In the engineering design process, students who lack relevant experience may encounter difficulty in using abstract concepts when attempting to develop solutions even though they have relevant knowledge. Prior to the implementation of engineering-focused STEM curricula, teachers should be aware of their students' current knowledge as well as the required knowledge and skills for the engineering design process so as to design appropriate activities. Meanwhile, teachers need to prepare to assess and respond to students' engineering design ideas and problems at appropriate moments (Johnson, Wendell, & Watkins, 2017). Because the time pressure induced by deadlines affects the manner in which students respond to a course, teachers should offer sufficient time to explore and experiment and ensure timely control of project progress, thereby preventing students from being perfunctory or falling behind schedule.

7.4.7 Knowledge of Learning Assessment

Engineering-focused STEM curricula aim to enable students to determine the effect of various realistic constraints on problem scenarios and potential solutions. Students can learn to apply mathematical knowledge and construct models for forecasting and analyzing the feasibility of strategies. Furthermore, embodying concepts through technology optimizes the design of final products. Such curricula focus on the learning process rather than merely final project presentations. Therefore, teachers must be capable of planning and implementing formative evaluations that assess students' learning outcomes through appropriate assessment tools.

7.5 Issues Related to Engineering-Focused STEM Curricula

Qualified teachers must be familiar with the characteristics and content of their teaching subjects as well as understand learner traits to determine why, what, and how to teach. However, STEM education should be based on actual and complex problem scenarios related to social context. Because excelling in a single domain is already fairly difficult for most teachers, expecting all teachers to have in-depth understanding of all STEM subjects is unreasonable. Berlin and White (2012) reported that teacher education for STEM curricula should provide teachers with opportunities to access similar, complementary, coordinated, conceptual, and procedural knowledge and skills, thus enabling teachers to gain a deep understanding of STEM content knowledge. Professional development for teachers of interdisciplinary education should be conducted by a group composed of teachers of various disciplines and grades (Love & Wells, 2018). Interdisciplinary teachers should discuss the most feasible curriculum based on school characteristics and resources, and the proposed integrated curriculum should mirror students' lives and learning needs (MacMath, 2011). Thus, strategic interdisciplinary collaboration increases the efficiency of integrated courses.

To facilitate the promotion of engineering-focused STEM curricula, this chapter proposed suggestions for the professional development of teachers, which may serve as a reference for educators wishing to enhance teachers' PCK. The following suggestions are proposed in addition to the aforementioned content and PCK needs.

1. Short-term business internships could give teachers a deeper understanding of engineering design operations. Teachers could acquire first-hand experience with the design and production process in the engineering sector and employ this experience in STEM courses.
2. STEM teachers should be innovative, work well in teams, and conduct self-directed learning.
3. STEM teachers should be at least familiar with one engineering-related discipline (e.g., mechanical engineering, electrical engineering, or architecture) as a basis for actively acquiring knowledge concerning technological development, which then can serve as a foundation for development and design of a curriculum.

Regarding the field of education, most previous discussions of PCK have explored a particular subject; they have rarely centered on STEM education as a whole. The perspectives of this chapter may serve as a reference for STEM education promotion as well as an appeal for more empirical research that can assist with the future development of teacher education.

References

American Association for the Advancement of Science. (1993). *Benchmarks for science literacy*. New York, NY: Oxford University Press.

Aranda, M. L., Lie, R., Selcen Guzey, S., Makarsu, M., Johnston, A., & Moore, T. J. (2018). Examining teacher talk in an engineering design-based science curricular unit. *Research in Science Education*, 1–19. https://doi.org/10.1007/s11165-018-9697-8.

Asunda, P. A. (2012). Standards for technological literacy and STEM education delivery through career and technical education programs. *Journal of Technology Education, 23*(2), 44–60.

Atman, C. J., Adams, R. S., Cardella, M. E., Turns, J., Mosborg, S., & Saleem, J. (2007). Engineering design processes: A comparison of students and expert practitioners. *Journal of Engineering Education, 96*(4), 359–379.

Berlin, D. F., & White, A. L. (2012). A longitudinal look at attitudes and perceptions related to the integration of mathematics, science, and technology education. *School Science and Mathematics, 112*(1), 20–30.

Berry, R. Q., III, Reed, P. A., Ritz, J. M., Lin, C. Y., Hsiung, S., & Frazier, W. (2004). Stem initiatives: Stimulating students to improve science and mathematics achievement. *Technology Teacher, 64*(4), 23–30.

Blackley, S., & Howell, J. (2015). A STEM narrative: 15 years in the making. *Australian Journal of Teacher Education, 40*(7), 102–112.

Budynas, R. G., & Nisbett, J. K. (2011). *Shigley's mechanical engineering design* (9th ed.). New York, NY: McGraw-Hill.

Business Roundtable. (2005). *Tapping America's potential: The education for innovation initiative.* Washington, DC: Author.

Bybee, R. W. (2013). *The case for STEM education: Challenges and opportunities.* Arlington, VA: NSTA Press.

Carr, R. L., Bennett, L. D., & Strobel, J. (2012). Engineering in the K-12 STEM standards of the 50 U.S. states: An analysis of presence and extent. *Journal of Engineering Education, 101*(3), 539–564.

Custer, R. L., & Daugherty, J. (2009). Professional development for teachers of engineering: Research and related activities. *The Bridge, 39*(3), 18–24.

De Miranda, M. A. (2017). Pedagogical content knowledge for technology education. In M. J. de Vries (Ed.), *Handbook of technology education* (pp. 685–698). Cham, Switzerland: Springer.

Fishman, B. J., Marx, R. W., Best, S., & Tal, R. T. (2003). Linking teacher and student learning to improve professional development in systemic reform. *Teaching and Teacher Education, 19*(6), 643–658.

Gess-Newsome, J., & Carlson J. (2013). *The PCK summit consensus model and definition of pedagogical content knowledge.* In the symposium Reports from the Pedagogical Content Knowledge (PCK) Summit, ESERA Conference, September 2013.

Helmes, J. V., & Stokes, L. (2013). *A meeting of minds around pedagogical content knowledge: Designing an international PCK summit for professional, community, and field development.* Retrieved from https://inverness-research.org/2014/09/22/ab_rpt_pck-summit/.

Herschbach, D. R. (2011). The STEM initiative: Constraints and challenges. *Journal of STEM Teacher Education, 48*(1), 96–122.

Householder, D. L., & Hailey, C. E. (Eds.). (2012). *Incorporating engineering design challenges into STEM courses.* Retrieved from ERIC database (ED537386).

International Technology and Engineering Educators Association. (2003). *Advancing excellence in technological literacy: Student assessment, professional development, and program standards.* Reston, VA: Author.

International Technology and Engineering Educators Association. (2007). *Standards for technological literacy: Content for the study of technology* (3rd ed.). Reston, VA: Author.

Johnson, A. W., Wendell, K. B., & Watkins, J. (2017). Examining experienced teachers' noticing of and responses to students' engineering. *Journal of Pre-College Engineering Education Research, 7*(1), 25–35. https://doi.org/10.7771/2157-9288.1162.

Kelley, T. R. (2010). Staking the claim for the 'T' in STEM. *Journal of Technology Studies, 36*(1), 2–11.

Kertil, M., & Gurel, C. (2016). Mathematical modeling: A bridge to STEM education. *International Journal of Education in Mathematics, Science and Technology, 4*(1), 44–55.

Lantz Jr., H. B. (2009). *Science, technology, engineering, and mathematics (STEM) education what form? What function?* Retrieved from https://dornsife.usc.edu/assets/sites/1/docs/jep/STEMEducationArticle.pdf.

Lewis, T. (2005). Coming to terms with engineering design as content. *Journal of Technology Education, 16*(2), 37–54.

Love, T. S., & Wells, J. G. (2018). Examining correlations between preparation experiences of US technology and engineering educators and their teaching of science content and practices.

International Journal of Technology and Design Education, 28(2), 395–416. https://doi.org/10. 1007/s10798-017-9395-2.

MacMath, S. L. (2011). *Teaching and learning in an integrated curriculum setting: A case study of classroom practices (Unpublished doctoral dissertation)*. Ontario, Canada: University of Toronto.

Magnusson, S., Krajcik, J., & Borko, H. (1999). Nature, sources, and development of pedagogical content knowledge for science teaching. In N. G. Lederman & J. Gess-Newsome (Eds.), *Examining pedagogical content knowledge* (pp. 95–132). Boston, MA: Kluwer/Springer.

Mentzer, N. (2011). High school engineering and technology education integration through design challenges. *Journal of STEM Teacher Education, 48*(2), 103–136.

Merrill, C., Custer, R. L., Daugherty, J., Westrick, M., & Zeng, Y. (2009). Delivering core engineering concepts to secondary level students. *Journal of Technology Education, 20*(1), 48–64.

Morrison, J., & Bartlett, R. (2009). STEM as curriculum: An experiential approach. *Education Week, 23*, 28–31.

National Research Council. (1996). *From analysis to action: Undergraduate education in science, mathematics, engineering, and technology*. Washington, DC: National Academies Press. https:// doi.org/10.17226/9128.

National Research Council. (2009). *Engineering in K–12 education: Understanding the status and improving the prospects*. Washington, DC: National Academies Press. https://doi.org/10.17226/ 12635.

National Research Council. (2011). *Successful K–12 STEM education: Identifying effective approaches in science, technology, engineering, and mathematics*. Washington, DC: National Academies Press. https://doi.org/10.17226/13158.

National Research Council. (2013). *Next generation science standards: For states, by states*. Washington, DC: National Academies Press. https://doi.org/10.17226/18290.

Park, S., & Oliver, J. S. (2008). Revisiting the conceptualization of pedagogical content knowledge (PCK): PCK as a conceptual tool to understand teachers as professionals. *Research in Science Education, 38*, 261–284.

Pinelli, T., & Haynie, W., III. (2010). A case for the nationwide inclusion of engineering in the K-12 curriculum via technology education. *Journal of Technology Education, 21*(2), 52–68.

Pitt, J. (2009). Blurring the boundaries: STEM education and education for sustainable development. *Design and Technology Education, 14*(1), 37–48.

Reimers, J. E., Farmer, C. L., & Klein-Gardner, S. S. (2015). An introduction to the standards for preparation and professional development for teachers of engineering. *Journal of Pre-College Engineering Education Research, 5*(1), 40–60. https://doi.org/10.7771/2157-9288.1107.

Saorín, J. L., Melián-Díaz, D., Bonnet, A., Carbonell Carrera, C., Meier, C., & De La Torre-Cantero, J. (2017). Makerspace teaching-learning environment to enhance creative competence in engineering students. *Thinking Skills and Creativity, 23*, 188–198. https://doi.org/10.1016/j. tsc.2017.01.004.

Shulman, L. S. (1986). Those who understand: Knowledge growth in teaching. *Educational Researcher, 15*(2), 4–14.

Shulman, L. S. (1987). Knowledge and teaching: Foundations of the new reform. *Harvard Educational Review, 57*(1), 1–23.

Sorenson, B. (2010). *Alaska S.T.E.M.: Education and the economy—Report on the need for improved science, technology, engineering and mathematics education in Alaska*. Juneau, AK: Juneau Economic Development Council. Retrieved from http://www.jedc.org/forms/ STEMEducationJEDCFinal.pdf.

Taylor, B. (2016). Evaluating the benefit of the Maker movement in K-12 STEM education. *Electronic International Journal of Education, Arts, and Science, 2*(1), 1–22.

Thomasian, J. (2011). *Building a science, technology, engineering, and math agenda: An update of state actions*. Washington, DC: National Governors Association Center for Best Practices. Retrieved from https://classic.nga.org/files/live/sites/NGA/files/pdf/1112STEMGUIDE.PDF.

Toulmin, C. N., & Groome, M. (2007). *Building a science, technology, engineering, and math agenda*. Washington, DC: National Governors Association. Retrieved from ERIC database (ED496324).

van Dijk, E. M., & Kattmann, U. (2007). A research model for the study of science teachers' PCK and improving teacher education. *Teaching and Teacher Education, 23*, 885–897.

Wells, J. G. (2013). Integrative STEM education at Virginia Tech: Graduate preparation for tomorrow's leaders. *Technology and Engineering Teacher, 72*(5), 28–35.

Chapter 8
Exploring the Affordances of Open-Source Sensors in Promoting Authenticity in Mathematics Learning

Joshua Lee Shanwei, Kenneth Y. T. Lim, Yuen Ming De and Ahmed Hilmy

8.1 Introduction

There is a growing recognition of the need for authenticity in mathematics education. Traditionally, mathematics lessons offer students inauthentic problems that do not show the relevance and usefulness of mathematics in real life or how to connect what students are taught in school to real-world issues, problems, and applications. The emphasis on science, technology, engineering, and mathematics (STEM) education in recent times could be perceived as a foundation and opportunity for innovation and change in mathematics classrooms. While there are multiple conceptions and approaches to STEM education, we emphasize authenticity and real-world situations as key factors in creating meaningful integration of the STEM disciplines and incorporation of real-world challenges and problems into the curriculum.

Mathematics education in Singapore has traditionally been based on arbitrary or fabricated contexts in rote learning, textbooks, and worksheets. In particular, the secondary school topic of data handling is still commonly taught through contexts in textbooks and worksheets that are often arbitrary to students. Where the data are not contrived, they are often static when presented in printed textbooks or preset worksheets, rendering them outdated or relatively non-localized at the point of student learning.

Lim's (2015) conception of disciplinary intuitions highlights the importance of learner intuitions as a basis of curricula design, evincing a need for educators to craft authentic learning experiences in class to surface and subsequently leverage these intuitions to promote deeper learning. Because of this disjunction between the contrived contexts and data, and the students' real-life experiences, learner intuitions are unlikely to be leveraged. Lave (1992) identified a source of difficulty in students' understanding of mathematics topics in a classroom setting as "the gulf between

J. L. Shanwei · K. Y. T. Lim (✉) · Y. Ming De · A. Hilmy
National Institute of Education, Nanyang Technological University, Singapore, Singapore
e-mail: kenneth.lim@nie.edu.sg

© Springer Nature Singapore Pte Ltd. 2019
Y.-S. Hsu and Y.-F. Yeh (eds.), *Asia-Pacific STEM Teaching Practices*,
https://doi.org/10.1007/978-981-15-0768-7_8

math learning in school and the everyday experience of children trying to bring their intuitions to bear while learning math in school" (p. 77). A lack of consideration for learner intuitions in formalized classroom instruction, typically emerging as inauthentic learning contexts, can result in disparities between what students intuitively know outside the classroom and what is formally taught inside the classroom (Cho & Hong, 2015).

Singaporean students' traditional and present lack of experience in working with real-world data starkly contrasts with the recognition of the importance of real-world contexts in mathematics learning for leveraging learner intuitions and intrinsic motivation in students. However, developments in the field of open-source sensors have given rise to inexpensive, modular sensors with straightforward assembly and localized setup that provide access to authentic, relevant, real-time data. Unlike other traditional data sources in the classroom, data collected by these sensors are real, localized, and current while still being easily retrieved, displayed, and worked on in the classroom. This provides the mathematics classroom with a source of real-world situations to spark students' interests and help them connect what they do in the classroom with their world around them.

8.2 Literature Review

Engaging and motivating students in mathematics lessons can be challenging. The traditional approach of *chalk and talk* can sometimes be problematic where the educator takes on the role of being the main source of the students' learning and of transmitting ideas, concepts, and information to the learner who is assumed to be a blank slate that requires the teacher's input (Burns & Brooks, 1970). The way mathematics is taught and the way it is traditionally represented in mathematics textbooks focuses primarily on procedural knowledge, that is, a sequence of steps for solving a type of problem or a set of rules and algorithms to be memorized. This representation encourages learners to suspend their existing understandings and prior concepts that may be relevant to their learning and subscribe to the "teacher-imposed methods of getting the correct answer" (Geist, 2010, p. 25). The emphasis on the need for the repetition of steps, the memorization of mathematical formulas, and the swiftness of solving problems can cause learners to perceive mathematics as a high-risk subject (Geist, 2010; Popham, 2008).

Moreover, this method of teaching gives rise to a view of mathematics as abstract and detached from reality. Even examples or story problems, which ask students to apply their knowledge of mathematics to solve problems that happen in a real-world scenario, "rarely invite students to use their common sense and their everyday experiences to solve those problems" (Popovic & Lederman, 2015, p. 129); but rather, questions such as figuring out when two trains will collide or how tall a lighthouse is based on the shadow it casts are often unrelatable to a young student.

There is an emerging shift from the traditional approach of teaching that emphasizes the learner's ability to recall disconnected facts and follow prescribed procedures and operations to the new teaching approach of facilitating "learning that enables critical thinking, flexible problem solving, and the transfer of skills and use of knowledge in new situations" (Darling-Hammond et al., 2008, p. 2). Even as we acknowledge that relating classroom mathematics to real-world contexts enhances students' understanding of mathematical ideas and motivates learning and applications (Gainsburg, 2008; Zeuli & Ben-Avie, 2003), how can schools and teachers, pedagogically and technically, connect the mathematical concepts they teach with their applications in real-world situations?

8.2.1 Stem

White (2014) suggested that the distinct disciplines of STEM were originally selected because these disciplines provide critical thinking skills that encourage students to solve problems within their area of study. These disciplines were originally ordered as science, mathematics, engineering, and technology—or SMET—that came from an initiative created by the National Science Foundation in the United States of America. The acronym was changed to STEM in 2001, and the term continued to grow in popularity and recognizability (Breiner, Harkness, Johnson, & Koehler, 2012). While there is no commonly agreed operational definition or conceptualization of STEM, most stakeholders seem to agree that STEM is about creating better teachers, students, and workforce in the twenty-first century (Breiner et al., 2012). Most people agree that the desired outcome of STEM is to prepare students at all levels with the skills necessary to compete in our rapidly advancing technological society. Allen, Webb, and Matthews (2016) adopted a view of STEM education as intentionally merging the separate disciplines in order to solve real-world problems. Therefore, a STEM curriculum is an integration of two or more STEM-related disciplines (inter-disciplinary) or a fusion of the four disciplines (transdisciplinary), which are used to help students acknowledge or respond to problems that exist in society.

Furthermore, Johnson (2013) argued that true integration of STEM education should be more than combining the subject areas, but specifically should be an instructional approach "which integrates the teaching of science and mathematics disciplines through the infusion of the practices of scientific inquiry, technological and engineering design, mathematical analysis, and 21st century interdisciplinary themes and skills" (p. 387). This definition deemphasizes content and emphasizes processes and skills, which set it apart from the traditional definitions of science and mathematics education.

The authors of this chapter view the implementation and use of technology that enables and facilitates the inquiry and investigation of real-world phenomena and data to solve real-world problems as a good starting point for implementing STEM education, specifically, in this case study of a traditional mathematical class structured into compartmentalized topics that did not rely on any authentic empirical data.

Kaiser (2002) cautioned "it is typical for German mathematics teaching that real-world examples discussed in lessons are not authentic real-world problems, but made to illustrate mathematical contents. Therefore, these examples give a quite artificial and far from reality impression" (p. 253). The introduction of authentic context and data into a traditional mathematics class can develop STEM thought processes and skills by solving real-world problems directly related to the learners.

8.2.2 Disciplinary Intuitions

Studies have suggested that conceptual knowledge (i.e., explicit or implicit understanding of the principles or relations) may have a greater influence on procedural knowledge than the reverse procedures (Rittle-Johnson & Alibali, 1999). Insufficient attention has been paid to the inter-functional dialectic relationship between scientific concepts and everyday concepts—disciplinary intuitions that build on ideas from cognitive development, sociolinguistics, and constructivism. Vygotsky (1978, p. 157) claimed that the development of everyday and scientific conceptions were two "parts of a single process: the development of concept formation which is affected by varying external and internal conditions but is essentially a unitary process, not a conflict of antagonistic, mutually exclusive forms of thinking." These tacit nascent understandings are disciplinary intuitions based on embryonic, unexamined, or unreinforced conceptions or misconceptions from lived experiences and early stages of learning that learners bring to the classroom and risk being dismissed by teachers who are working within curricular constraints, believing that misconceptions can to be quickly set right by the formal curriculum.

Lim (2015) argued that, when teachers and learners are afforded opportunities to acknowledge, surface, and negotiate these disciplinary intuitions, it will result in much more enduring understandings on the part of the learners. The Disciplinary Intuitions Model (Fig. 8.1) is based on the belief that learning is predicated upon a person's lived experiences. Learning that does not build upon intuitions from lived experiences is purported to become purely theoretical information that will be difficult for students to apply in the real world. Writing specifically about mathematics word problems used in schools, Lave (1992) observed that:

> Children's intuitions about the everyday world are in fact constantly violated in situations in which they are asked to solve word problems. This discontinuity by itself may help create the division between 'real' and 'other' mathematics by conveying the message that what the children know about the real world is not valid. (p. 78)

Students who are unable to connect new knowledge with knowledge and concepts that they already tacitly or explicitly know—due to an inability to connect the stimulus in the instructional setting to the performance of the task in a generalization setting—may meaninglessly relegate the new stimulus to the theoretical in isolation from their intuitions. Unable to construct new meanings with these new stimuli, students may find understanding or remembering such stimuli difficult and may find it harder still to

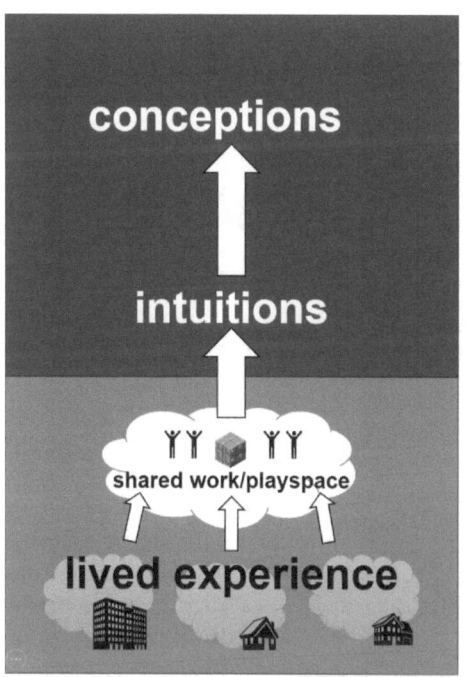

explicit.
- most curriculum approaches focus here

tacit.
- beyond the control of traditional curriculum / instruction
- immersive environments and sandboxes as a shared proxy lived experience

Disciplinary Intuitions

Fig. 8.1 Disciplinary Intuitions Model of learners' cognition in the learning process

apply them to problems in the real world. Therefore, educators and their lessons need to be informed not only by learners' explicit conceptions and misconceptions but also by a further understanding of the tacit intuition learners bring to their learning. This requires educators to construct learning environments that can surface and leverage these learner intuitions. A carefully designed, shared work–play space can increase the likelihood of the students making a connection between their lived experiences and the learning context so that their intuitions from lived experiences can be engaged in their thinking and learning processes. Educators will then be able to identify and build upon these learner intuitions in class for deeper learning and address them while they are still malleable.

8.2.3   Authentic Learning

Reeves, Herrington, and Oliver (2002) suggested that authentic learning is a student-centric pedagogy whereby learning takes place in a crafted learning environment that is situated in or simulates the real world. Students construct understanding based on their past experiences, prior knowledge, current experiences, and sociocultural factors involving real-world problems with practical, tangible solutions (Lombardi, 2007). The realism of this learning environment allows direct connections to be made by students between their lived experiences and the learning taking place in the crafted learning environment or context. Therefore, learner intuitions can be leveraged by setting up authentic learning environments.

Reeves et al. (2002) proposed a ten-element checklist that characterizes the authentic learning experience. Lombardi (2007) explained and elaborated upon the individual items and suggested how they may be adapted to any subject matter domain (Table 8.1). The central theme running through most of these design elements is real-world relevance in the learning process. Hence, using this list of design elements as a checklist in the crafting of a lesson plan or learning package, educators can create an authentic learning context that is as close to the real world as possible so that students can relate lived experiences to the given learning context.

8.3 Case Study

Despite the growing rhetoric about STEM education, what this means for teaching and learning in mathematics education is still largely under-conceptualized. This research used a case study designed to explore and document a learning experience in a secondary mathematics classroom that took into account STEM interdisciplinary and real-life context, as well as disciplinary intuitions and the authentic learning framework. The classroom activities and analyses were designed by the students' regular mathematics teachers, who had several discussions with the project team. The research team's primary role was to assist in data collection and to conduct workshops during the school's regular personal development hours to introduce the hardware and learning frameworks to the teachers.

8.3.1 Context

This set of lessons was designed and conducted with classes at Yio Chu Kang Secondary School (YCKSS), a coeducational government school in Singapore, in conjunction with the National Institute of Education, an autonomous institute of Nanyang Technological University in Singapore. The intervention incorporated the use of data collected by open-source sensors into the mathematics curriculum of the school. The purpose of the instructional unit was to let students work with and learn from localized data, thereby allowing them to practice their skills with real-world data collected from their local environment. Through this intervention, this study aimed to investigate the following research question: Can the use of real-time localized data enable the construction of an authentic STEM learning context in a mathematics classroom?

Table 8.1 Ten design elements of authentic learning (adapted from Lombardi, 2007)

Design element	Description
1. Real-world relevance	Authentic activities match the real-world tasks of professionals in practice as nearly as possible. Learning rises to the level of authenticity when it asks students to work actively with abstract concepts, facts, and formulae inside a realistic and highly social context
2. Ill-defined problem	Challenges cannot be solved easily by the application of an existing algorithm. Instead, authentic activities are relatively undefined and open to multiple interpretations
3. Sustained investigation	Problems cannot be solved in a matter of minutes or even hours. Instead, authentic activities comprise complex tasks to be investigated by students over a sustained period of time, requiring a significant investment of time and intellectual resources.
4. Multiple and extraneous resources	Authentic activities provide the opportunity for students to examine the task from a variety of theoretical and practical perspectives, using a variety of resources, and requires students to distinguish relevant from irrelevant information in the process
5. Collaboration	Success is not achievable by an individual learner working alone. Authentic activities make collaboration integral to the task, both within the course and in the real world
6. Reflection (metacognition)	Authentic activities enable learners to make choices and reflect on their learning, both individually and as a team or community
7. Interdisciplinary perspective	Relevance is not confined to a single domain or subject matter specialization. Instead, authentic activities have consequences that extend beyond a particular discipline, encouraging students to adopt diverse roles and think in interdisciplinary terms
8. Integrated assessment	Assessment is not merely summative in authentic activities but is woven seamlessly into the major task in a manner that reflects real-world evaluation processes
9. Polished products	Activities culminate in the creation of a whole product rather than an exercise or sub-step in preparation for something else
10. Multiple interpretations and outcomes	Activities allow a range and diversity of outcomes open to multiple solutions of an original nature, rather than a single correct response obtained by the application of rules and procedures

8.3.2 Students' Needs

YCKSS mathematics teachers began by identifying students' needs and challenges in learning data handling skills and content. The teachers identified three areas that could be improved by a real-world, data-driven intervention: mathematical skills related to the topic of mean, median, and mode; reasoning, communication, and connections; and mathematical beliefs, interest, and motivation.

The teachers highlighted that conventionally, while their students may acquire a theoretical understanding of what the different measures of central tendency are and how to obtain them quantitatively, they were often unable to arrive at a clear qualitative understanding of the pros and cons in using different measures or were unable to articulate their understanding clearly. One common feedback the teachers received is that secondary school mathematics is not relevant to the student's life, which they believe negatively affects student motivation in practicing mathematics.

8.3.3 Pedagogical Considerations

The teachers had several meetings to brainstorm on how to design a STEM learning environment using data in the mathematics classroom. They decided to base their design on inquiry-based learning (Pedaste et al., 2015). The three-phase lesson focused on solving a central problem using a teacher-guided, student-directed, investigative inquiry-based activity consisting of:

1. Orientation and Conceptualization: Context, trigger, or question that sets up the central problem to be solved, along with hypothesis generation.
2. Investigation: A systematic method to gather and process relevant information through data or otherwise to solve the problem.
3. Conclusion: Synthesis of the findings to arrive at a conclusion or product in response to the problem.

The foremost task was to come up with a trigger inquiry question that students are likely to find interesting or relevant to them and can be answered through investigation of current, localized, real-world data. The question selected was: *Where is the best spot to study in school?* This question is relevant and understandable to the students, who will also have diverse personal experiences in deciding where in the school to go and do their work. The context was specified to after-school self-study, during the time period of 3–6 p.m., in any location within the school. The teachers decided on six different environmental factors that may be relevant: temperature, humidity, visible light, infrared radiation, ultraviolet radiation, and noise level across the possible locations to determine the suitability of study. The locations will be determined by the students in class. The resulting data will be analyzed and processed by the students so that they could compare the factors across different locations and draw logical conclusions. The goal was for the students to mathematically process the data and use the result to logically deduce their answer to the trigger inquiry question. A lesson worksheet was crafted for the students that would allow them to collectively record their activities and to express themselves in their own words.

A key tenet of a good investigative inquiry-based activity is the personal relevance of the trigger to the students. Ryan (1995) and Ryan and Deci (2000a, 2000b) argued that an increase in students' perceived value of or interest in an activity moves students further along the self-determination continuum toward being more motivated both extrinsically and internally. Ormrod (2008) also believed this emotional engagement positively influences the degree to which students choose to engage in classroom activities. An investigative inquiry-based activity such as this that has real, personal relevance fulfills many of the design elements in authentic learning.

- Real-world Relevance: The activities are in fact real-world questions that the students find themselves answering every day. It stems from a real need found in the students' environment.
- Ill-defined Problem: Although the problem is one that is easily understood by the student, there are various factors that need to be considered and the students need to evaluate their findings to respond meaningfully to the problem. Moreover, the result is open to debate and other considerations.
- Sustained Investigation: The data are collected throughout a week during the school term, providing the students with variation and trends across different days.
- Multiple and Extraneous Resources: Students are given extraneous real-world data and must contemplate the relevancy of each datum.
- Collaboration: This activity is designed to be a group investigative task; the group needs to work together, discuss, and decide on an answer.
- Metacognition: Reflective learning is fostered through individual or group prompts by the teacher in class or in the guiding questions in the worksheet.
- Multiple Interpretations and Outcomes: Solutions were not evaluated as dichotomous (right or wrong). The activities allow for competing solutions with nuanced evaluations and diverse interpretations of outcomes.

8.3.4   Technical Affordances

The weather sensor network set up at the school consisted of five sensor motes—each placed in a different part of the school—that transmitted data via wireless radio to an Internet-connected hub device, which then uploaded the data to a cloud service (i.e., Google Sheets). Teachers and students could access the cloud data to obtain the sensor data remotely via their smart devices (e.g., computer, tablet, or mobile phone).

Each sensor mote was constructed from the following components:

- Arduino UNO Microcontroller Board: An open-source, programmable electronics platform that is cheap, easily accessible, and supported by a wide ecosystem of compatible extension parts.

- XBee Radio Module: A medium-range, programmable radio module that can be used to transmit data between devices up to 100 m away from each other.
- Temperature and Humidity Sensor: Integrated temperature and humidity sensor that measures temperature in degrees celsius and relative humidity as a percentage.
- Sunlight Sensor: Measures visible light in lumens, infrared light in lumens, and ultraviolet light in UV index.
- Noise Sensor: Measures noise levels.
- 5000 mAh Power Bank (portable charger): Arduino UNO microcontroller board was powered by a power bank that could keep the device running for 5 days.

The hub device was constructed from the following components:

- Raspberry Pi Single-board Computer: A portable computing device with built-in storage that runs an open-source operating system (Linux).
- XBee Radio Module: Same as the one attached to each sensor mote, but this one was programmed to send and receive data from all five sensor motes.

8.3.5 Implementation

The teachers designed a 2-week lesson package for implementation during the regular Secondary Mathematics 2 classes. The lesson package comprises a lesson plan for teacher instructions and student activities and an accompanying student worksheet to stimulate reflective thinking and consolidate learning.

The lesson package was carried out in two classes: the first was from February 12–21, 2018, and the second was one month later from March 12–22. The seeming long duration of each lesson package includes the week-long data collection by the sensors as well as classroom instruction time. After the data collection was completed, the students received and worked on the compiled data in their mathematics classes. These lessons were observed by the research team.

The first lesson for each class introduced the investigative task of determining the best study spot in the school based on environmental factors that would be conducive to study. The students were divided into groups to study the relevant factors and formulate their hypothesis on where the best location may be. At the end of the first lesson, each group submitted a few locations they would like to investigate. Although the teachers did not assign any specific locations, groups were encouraged to choose different places to investigate. Some examples of the locations chosen were a canteen or benches near the general office on the ground floor, study tables near classrooms or the elevator on the second floor, and study tables on the third and fourth floors. The sensors hardware was packed into transparent lunchboxes and cable-tied to fixtures in those locations (Fig. 8.2). The data were wirelessly collected for a week and stored on spreadsheets on the cloud storage service.

Six different environmental readings were captured by the sensors: temperature, humidity, visible light, infrared radiation, ultraviolet radiation, and sound or noise level. Each sensor was set to record one set of readings every five minutes throughout the week for a specified location. These data were formatted and printed for use by the students during class time. Each group was given a printout of the data collected for each location they picked (Fig. 8.3).

With the collected data, the teacher conducted lessons on data processing and analysis, using the mathematical skills the students practiced in class. The teacher reintroduced the investigation topic and explained how the data were generated. Each group was given their relevant data printout; every student must complete an individual worksheet (Figs. 8.4 and 8.5). The teacher facilitated each group's discussion and data handling processes by going around the class to engage students in Socratic questioning and by responding to queries. Class discussions were then conducted to consolidate, compare, and evaluate group results.

Fig. 8.2 Sensors set up in various locations that students wanted to investigate

128   J. L. Shanwei et al.

Sensor 1: _____

Time	Temperature (°C)	Humidity (%)	Visible light (lm)	Infrared light (lm)	Ultraviolet light (UV Index)	Sound (range 0-1023)
Wed, 21/2/18, 15:00	28.83	94.33	256.83	286.78	0.03	31.50
Wed, 21/2/18, 15:05	29.00	95.00	256.00	280.45	0.04	28.00
Wed, 21/2/18, 15:10	29.00	95.00	256.00	277.33	0.03	32.73
Wed, 21/2/18, 15:15	29.00	95.00	256.00	276.50	0.03	28.75
Wed, 21/2/18, 15:20	29.00	95.00	256.00	276.00	0.03	25.00
Wed, 21/2/18, 15:25	29.00	95.00	257.83	296.00	0.04	30.67
Wed, 21/2/18, 15:30	29.00	95.00	258.90	295.90	0.04	31.90
Wed, 21/2/18, 15:35	29.00	95.00	271.00	402.00	0.08	47.00
Wed, 21/2/18, 15:40	0.00	0.00	0.00	0.00	0.00	0.00
Wed, 21/2/18, 15:45	29.00	95.00	256.00	282.50	0.04	46.17
Wed, 21/2/18, 15:50	29.00	95.00	267.00	330.67	0.05	78.67
Wed, 21/2/18, 15:55	29.00	95.00	257.17	281.17	0.03	32.08
Wed, 21/2/18, 16:00	29.00	95.00	257.40	292.60	0.04	38.20
Wed, 21/2/18, 16:05	29.00	95.00	267.50	378.83	0.07	49.83
Wed, 21/2/18, 16:10	29.00	95.00	256.00	283.75	0.03	31.25
Wed, 21/2/18, 16:15	29.00	95.00	271.00	411.00	0.08	27.00
Wed, 21/2/18, 16:20	29.00	95.00	270.50	412.00	0.08	43.00
Wed, 21/2/18, 16:25	29.77	95.00	271.46	407.00	0.08	30.54
Wed, 21/2/18, 16:30	30.00	95.00	270.86	402.57	0.08	25.29
Wed, 21/2/18, 16:35	30.00	95.00	270.67	400.33	0.08	51.67
Wed, 21/2/18, 16:40	30.00	95.00	271.14	395.86	0.08	37.71
Wed, 21/2/18, 16:45	30.00	95.00	264.22	339.67	0.05	28.33
Wed, 21/2/18, 16:50	29.63	95.00	261.88	316.75	0.05	29.63
Wed, 21/2/18, 16:55	29.25	95.00	268.00	344.00	0.06	20.75

Fig. 8.3 Datasheet distributed to students for the lesson

8.4 Observations

Two separate group interviews were held—one after each class completed the entire lesson. The aim of the interviews was to elicit students' perspectives on their learning experiences and their opinions on the curriculum. Students' worksheets were collated and analyzed with the aid of the participating teacher to determine if students had engaged meaningfully with the exercises.

Findings

5. Record your assigned location: _Outside Staffroom_

6. Record the raw data in the table below:

Day ~~Fri~~ 4 3

Time	temperature/°C	brightness/lumen	noise level
15:00	31.00	262.33	473.11
15:10	31.00	261.94	520.47
15:20	30.00	262.10	524.80
15:30	29.00	261.76	484.35
15:40	29.00	261.24	418.88
15:50	28:58	261.47	436.95
16:00	28.00	260.88	456.35
16:10	28.00	261.44	481.19
16:20	28.00	261.94	494.12
16:30	28.00	262.00	504.76
16:40	28.00	261.74	574.26
16:50	28.00	261.29	521.82
17:00	27.13	261.47	507.53
17:10	27.00	261.00	518.71
17:20	27.00	261.00	521.71
17:30	27.00	261.47	527.82
17:40	27.00	261.78	523.33
17:50	27.00	261.78	521.28

7. What was your time interval when you record the data in the table in 6? Explain your rationale.

10 mins It is convenient to fill in the data and the mean will be more accurate

Data Processing

8. From the data collected from the sensors, find the mean temperature in °C, brightness in lumen, and noise level recorded between 3.00 -6.00 pm over a week (Monday to Friday) for one location. Record the data in the table below.

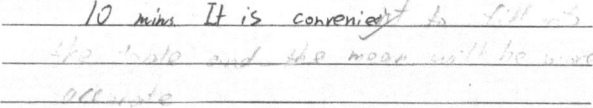

Time	mean temperature/°C	mean brightness/lumen	mean noise level
Day 1	31.98	266.98	469.16
Day 2	33.32	250.30	453.79
Day 3	26.76	276.12	468.32
Day 4	31.62	266.34	300.52
Day 5			

3

Fig. 8.4 A part of the student's worksheet involving data analysis and processing

Your class is tasked to find out which location in Yio Chu Kang Secondary School, excluding the library, is most conducive for self-study.

To determine the most conducive location, we will consider the following 3 factors
   (1) Brightness,
   (2) Noise level,
   (3) Temperature.

In groups of 5, each group will collect data on the **above factors** at **one** location in the school and study how each factor **varies over one week**.
Each group will then **analyse** the data to **determine** whether that location is suitable for self-study.

4.   List 5 locations that you think are suitable for self-study.

   • Canteen
   • Outside GO
   • Outside staffroom
   • leave 3 study area
   • level 4 study area

2

Fig. 8.5 A student draws from personal experiences and intuitions to hypothesize a possible solution

8.4.1 Students Appreciated the Use of Relatable Problems and Real Data

A: Fun because we get to like play [and] use the real-life data instead of the textbook … so like our knowledge about the school surroundings is better.

B: It just actually shows like … what the people from outside are actually dealing with … because now in the textbook is more simplified … but when we grow older or get … jobs maybe we have to deal with these kinds of things or maybe harder.

During the group interviews, students were asked what the differences are, if any, between doing a mathematics lesson using localized data from their school, compared to using arbitrary textbook data, or data collected from somewhere else in the world. Of the students interviewed, the majority expressed they liked this lesson similarly with localized data from their school because it is relevant to them.

Students also shared that they felt a disjunction between the real-world context given in class and the actual process of data-collection outside of mathematics class but within their school. Even though they understood that the data presented in class were collected by sensors during the test period at their chosen venues, students expressed their wish to be more directly and personally involved in the sensors' setup and data collection process. They wanted to understand how the hardware of the sensors resulted in the digital data in the spreadsheets.

8.4.2 Students Connected Mathematics to Everyday Experiences

When asked to relate how different temperatures, brightness, and noise levels affect their study environment, many students drew on their own personal lived experiences to describe how the environmental factors affected them (Fig. 8.6).

The students gave different answers when asked to estimate an acceptable data range for a conducive study environment. As seen in Fig. 8.7, these two students held quite different views on the acceptable range on three factors. The questions and activity were constructed to connect mathematics topics to experiences, situations and contexts outside of the mathematics classroom, including students' own experiences and understandings of the world, which is a critical element important to student learning (Aguirre et al., 2013; Turner & Font, 2007). Connecting school mathematics with everyday experiences can serve as valuable resources for learning (Civil, 2002, 2007) because realistic contexts and situations will enhance concepts and skills.

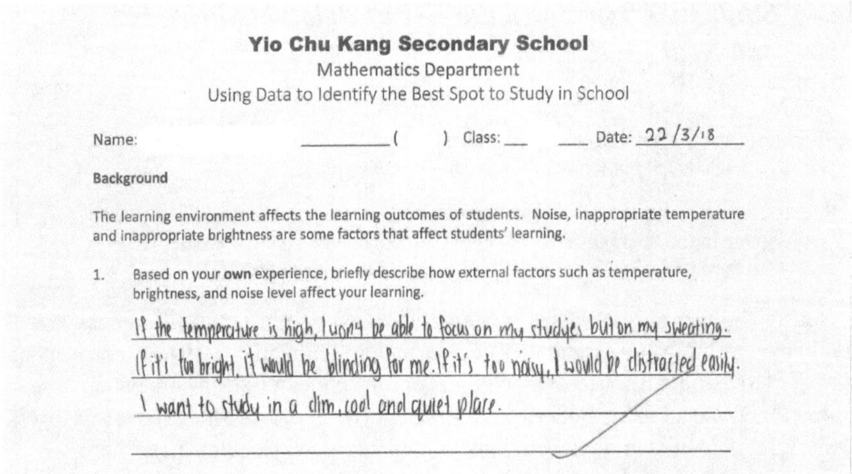

Yio Chu Kang Secondary School
Mathematics Department
Using Data to Identify the Best Spot to Study in School

Name: _____ ( ) Class: __ __ Date: _22_/3/18_

Background

The learning environment affects the learning outcomes of students. Noise, inappropriate temperature and inappropriate brightness are some factors that affect students' learning.

1. Based on your **own** experience, briefly describe how external factors such as temperature, brightness, and noise level affect your learning.

    If the temperature is high, I won't be able to focus on my studies but on my sweating. If it's too bright, it would be blinding for me. If it's too noisy, I would be distracted easily. I want to study in a dim, cool and quiet place.

Fig. 8.6 One example of a student responding to the worksheet with personal experiences

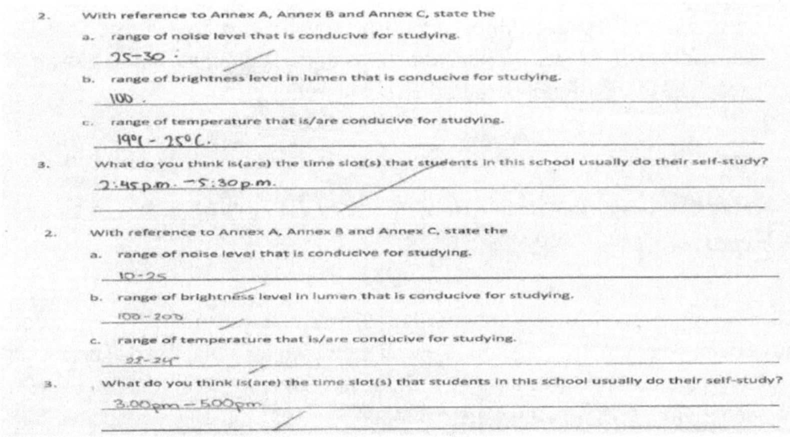

2. With reference to Annex A, Annex B and Annex C, state the
    a. range of noise level that is conducive for studying.
        25-30 :
    b. range of brightness level in lumen that is conducive for studying.
        100
    c. range of temperature that is/are conducive for studying.
        19°C - 25°C.
3. What do you think is(are) the time slot(s) that students in this school usually do their self-study?
    2:45 p.m. - 5:30 p.m.

2. With reference to Annex A, Annex B and Annex C, state the
    a. range of noise level that is conducive for studying.
        10-25
    b. range of brightness level in lumen that is conducive for studying.
        100-200
    c. range of temperature that is/are conducive for studying.
        23-24°
3. What do you think is(are) the time slot(s) that students in this school usually do their self-study?
    3.00 pm - 5.00 pm

Fig. 8.7 Comparison between two students with different views on the acceptable range of the quantities of three factors

8.4.3 Students Questioned the Textbook and Analyzed Possible Solutions to Real-World Problems

There was an unexpected technical error in the setup that resulted in a 5-min period wherein no data were recorded by the sensors or received by the cloud. The teacher skillfully recognized this as a teachable moment and expounded to students how zero-readings reflect the reality of imperfections in the real-world data collection, versus textbook data that are fabricated and *too perfect*. This presented the students with a dilemma. For example, the formula for calculating mean is to add all the values then divide by the number of values. But what do they do when one row in the datasheet is a series of zeros due to the technical error. The students were unsure if they should include those zeros in the calculation of their means, medians, and modes. When trying to calculate the means, some groups were unable to decide if they should "divide by 5 without the zeros or 6 to include the zeros." Unable to come to a group consensus, these groups turned to the teacher for help.

These student actions showed that they had managed to apply their understanding of the lesson to an unexpected challenge presented by the primary data. They would not have been able to correctly question the number of readings to incorporate in their calculation had they not understood that the principle of calculating the mean of a set of data equals to the value of all the data points divided by the number of data points. More importantly, by discussing the validity of data points, they made the leap from following formulas and instructions in their mathematics textbook to considering the fuzzy, real-world problem of data validity.

In this example, the teacher recognized that the students' reasoning for not including the zero-readings demonstrated their understanding of what are meaningful data and their ability to make a decision on rejecting data that do not help in interpreting the results (Fig. 8.8). The mathematics class broadened from one of the applying formulas to numbers in textbooks to a discussion on data analysis. One student described it like this: "We will be losing another data (point), so the average will be lesser in some way, it's not accurate." Another student verbally explained it as: "The mean will drop down, because it's a zero …, it's a huge difference."

Some groups wanted the teacher to provide an answer to what should be done with regard to the zero-readings; however, the teacher skillfully challenged the students, asking them to first try to explain what they thought they should do and why. We observed that the students were experiencing a real-life authentic problem in a mathematics lesson and were coming up with their own solutions, adapting their behaviors, and making meaningful decisions. They did this by utilizing and manipulating their prior knowledge, drawing upon their experiences and knowledge to guide their choices and to help them determine their next steps within the context of the problem. We find this to be a prime example of authentic learning in a STEM lesson, which is equipping students with essential life skills.

8.5 Discussion

The results of this study support the hypothesis that open-source sensors enabled the construction of an authentic STEM learning context that can be used in learning environments to affect deeper mathematical and problem-solving skills. We observed that students broke down the main question into different parts, sustained working with the data collected, rejected extraneous data like ultraviolet and infrared light readings, collaborated in groups, and came up with multiple interpretations and

Findings

5.   Record your assigned location: _Level 4 study area_

6.   Record the raw data in the table below:

Day 1

Time	temperature/°C	brightness/lumen	noise level
3.00	31.00	256.00	26.00
3.30	0.00	0.00	0.00
4.00	31.00	256.00	25.00
4.30	0.00	0.00	0.00
5.00	0.00	0.00	0.00
5.30	0.00	0.00	0.00
5.55	30.00	256.00	29.00

Data Processing

8.   From the data collected from the sensors, find the mean temperature in °C, brightness in lumen, and noise level recorded between 3.00–6.00 pm over a week (Monday to Friday) for one location. Record the data in the table below.

Time	mean temperature/°C	mean brightness/lumen	mean noise level
Day 1	30.64	256	26.67
Day 2	—	—	—
Day 3	—	—	—
Day 4	—	—	—
Day 5	—	—	—

Fig. 8.8 Example of a student's work showing his/her choice in ignoring the zero-readings in the calculation of the means

conclusions from the data collected. When asked in class and during the interviews, several students could express why they felt the task had real-world relevance to them.

Some students' first impression of the data sheet was that there were too many overwhelming numbers on the paper. This problem can be exacerbating for students who lack confidence or students with learning difficulties like dyslexia. During the interview, one student expressed a preference for the teaching of averages using textbook data instead. This student may represent others who feel more comfortable with traditional textbook examples, with less and often simplified data that are easier to understand, or who may be weaker and less confident in their ability to handle larger data sets. More research and lesson design can be conducted to engage these learners and bring about their persistence and eagerness to engage in mathematical challenges.

The students considered the lesson enjoyable and meaningful. These students were curious and cared about their school environment. They were motivated to discover and improve the schools for themselves and others at the school. Rather than studying a hypothetical problem in a textbook, they were examining a problem that mattered to them, and this led to relevance in their learning. In fact, they expressed an interest to get involved in the data collection and related activities—setting up hardware, determination of locations, problems to investigate, etc.

This led to a discussion of a shortcoming in this case study—the omission of one of the key design elements in authentic learning and a common component of STEM education—the interdisciplinary perspective. Lessons at YCKSS, and Singapore in general, are largely based on the conventional isolation of disciplines within established school systems that do not yet allow for cross-disciplinary or multidisciplinary learning, much less interdisciplinary learning. Lombardi (2007) asserted that learning should take place across multiple disciplines in an interdisciplinary way because it mirrors reality and real-world tasks that professionals encounter. Lessons designed around the use of open-source hardware sensors are well suited for interdisciplinary learning. For example, science and geography lessons can be designed to investigate and monitor the local environment, the data can be processed and visualized in mathematics lessons, the construction of suitable casings for the sensor can be a design and technology task, and the coding of the hardware can be an exercise in coding lessons.

To make the tasks even more meaningful, we can construct a scenario that allows students to create a useful, shared outcome. For example, the schools might set up a study corner in the locations that the student determined is conducive to self-study or building a model of the new study area to show changes they were planning to implement, or perhaps the task can be to monitor the school garden or pond to improve plant or fish growth.

8.6 Conclusion

Experiences that do not encourage children to make meaning from their learning will quickly be forgotten. Any learning experience should aim to instill authenticity into every task, lesson, and unit to ensure that students are able to develop real-world, problem-solving skills. Thanks to the emergence of a new set of technological tools, we can offer students more authentic learning experiences. This study shows that today's technology can enable the construction of an authentic STEM learning context that can be used to surface and leverage learner intuitions for deeper mathematical learning. This case study is a localized study, crafted in conjunction with a secondary school to meet its students' needs. We think that there is the potential to involve multiple departments to make the learning task interdisciplinary and even more authentic for students. We hope that school systems can allow for such lessons in future sensor-based interventions. Nevertheless, this work contributes to affirming the use of open-source sensors in a myriad of STEM learning environments to target student needs in schools today. With such a vast space of pedagogical design before us, it is hoped that this work can inform the possible merits and caveats of this STEM pedagogy, inspire teachers and academics on related sensor-based interventions, and help educators be more cognizant of the roots of the conceptions that students bring to learning environments.

Acknowledgements The authors would like to express their gratitude to the following persons for their invaluable part in this research project: Mdm. Goh Shwu Jun, Mr. Jeremy Chen, and Mdm. Lee Ching Fong from YCKSS for their collaboration and in shaping shared learning, and to Johnervan and Jonathan Lee for their continual encouragement and help throughout the project. The authors acknowledge the funding support for this project from Nanyang Technological University under the Undergraduate Research Experience on CAmpus (URECA) program.

References

Aguirre, J. M., Turner, E. E., Bartell, T. G., Kalinec-Craig, C., Foote, M. Q., Roth McDuffie, A., et al. (2013). Making connections in practice: How prospective elementary teachers connect to children's mathematical thinking and community funds of knowledge in mathematics instruction. *Journal of Teacher Education, 64*(2), 178–192.

Allen, M., Webb, A. W., & Matthews, C. E. (2016). Adaptive teaching in STEM: Characteristics for effectiveness. *Theory into Practice, 55*(3), 217–224.

Breiner, J. M., Harkness, S. S., Johnson, C. C., & Koehler, C. M. (2012). What is STEM? A discussion about conceptions of STEM in education and partnerships. *School Science and Mathematics, 112*(1), 3–11.

Burns, R. W., & Brooks, G. D. (Eds.). (1970). *Curriculum design in a changing society.* Englewood Cliffs, NJ: Educational Technology Publications.

Cho, Y. H., & Hong, S. Y. (2015). Mathematical intuition and storytelling for meaningful learning. In K. Y. T. Lim (Ed.), *Disciplinary intuitions and the design of learning environments* (pp. 155–168). Singapore: Springer. https://doi.org/10.1007/978-981-287-182-4_12.

Civil, M. (2002). Culture and mathematics: A community approach. *Journal of Intercultural Studies, 23*(2), 133–148.

Civil, M. (2007). Building on community knowledge: An avenue to equity in mathematics education. In N. S. Nasir & P. Cobb (Eds.), *Improving access to mathematics: Diversity and equity in the classroom* (pp. 105–117). New York, NY: Teachers College Press.

Darling-Hammond, L., Barron, B., Pearson, P. D., Schoenfeld, A. H., Zimmerman, T., Cervettti, G., et al. (2008). *Powerful learning: What we know about teaching for understanding.* San Francisco, CA: Jossey-Bass.

Gainsburg, J. (2008). Real-world connections in secondary mathematics teaching. *Journal of Mathematics Teacher Education, 11*(3), 199–219.

Geist, E. (2010). The anti-anxiety curriculum: Combating math anxiety in the classroom. *Journal of Instructional Psychology, 37*(1), 24–31.

Johnson, C. C. (2013). Conceptualizing integrated STEM education. *School Science and Mathematics, 8*(113), 367–368.

Kaiser, G. (2002). Educational philosophies and their influence on mathematics education—An ethnographic study in English and German mathematics classrooms. *Zentralblatt für Didaktik der Mathematik, 34*(6), 241–257.

Lave, J. (1992). *Word problems: A microcosm of theories of learning* (pp. 74–92). Context and Cognition: Ways of Learning and Knowing.

Lim, K. Y. T. (Ed.). (2015). *Disciplinary intuitions and the design of learning environments.* Singapore: Springer.

Lombardi, M. M. (2007). Authentic learning for the 21st century: An overview. *EDUCAUSE Learning Initiative, 1,* 1–12.

Ministry of Education. (2012). *Mathematics syllabus secondary one to four express course normal (academic) course.* Singapore: Author. Retrieved from https://www.moe.gov.sg/docs/default-source/document/education/syllabuses/sciences/files/mathematics-syllabus-sec-1-to-4-express-n(a)-course.pdf.

Ormrod, J. E. (2008). *Educational psychology: Developing learners* (6th ed.). Upper Saddle River, NJ: Pearson.

Pedaste, M., Mäeots, M., Siiman, L. A., de Jong, T., van Riesen, S. A., Kamp, E. T., ... & Tsourlidaki, E. (2015). Review: Phases of inquiry-based learning: Definitions and the inquiry cycle. *Educational Research Review,* 1447–1461. doi:https://doi.org/10.1016/j.edurev.2015.02.003.

Popham, W. J. (2008). Timed test for tykes? *Educational Leadership, 65*(8), 86–87.

Popovic, G., & Lederman, J. S. (2015). Implications of informal education experiences for mathematics teachers' ability to make connections beyond formal classroom. *School Science and Mathematics, 115*(3), 129–140.

Reeves, T. C., Herrington, J., & Oliver, R. (2002). Authentic activities and online learning. In T. Herrington (Ed.), *Quality conversations: Research and development in higher education* (Vol. 25, pp. 562–567). Jamison, ACT: Higher Education Research and Development Society of Australasia.

Rittle-Johnson, B., & Alibali, M. W. (1999). Conceptual and procedural knowledge of mathematics: Does one lead to the other? *Journal of Educational Psychology, 91,* 175–189.

Ryan, R. M. (1995). Psychological needs and the facilitation of integrative processes. *Journal of Personality, 63*(3), 397–427.

Ryan, R. M., & Deci, E. L. (2000a). Intrinsic and extrinsic motivations: Classic definitions and new directions. *Contemporary Educational Psychology, 25,* 54–67.

Ryan, R. M., & Deci, E. L. (2000b). Self-determination theory and the facilitation of intrinsic motivation, social development, and well-being. *American Psychologist, 55*(1), 68–78.

Turner, E. E., & Font, B. T. (2007). Problem posing that makes a difference: Students posing and investigating mathematical problems related to overcrowding at their school. *Teaching Children Mathematics, 13*(9), 457–463.

Vygotsky, L. S. (1978). *Mind in society: The development of higher psychological processes.* Cambridge, MA: Harvard University Press.

White, D. W. (2014). What is STEM education and why is it important? *Florida Association of Teacher Educators Journal, 1*(14), 1–9.

Zeuli, J. S., & Ben-Avie, M. (2003). Connecting with students on a social and emotional level through in-depth discussions of mathematics. In N. M. Haynes, M. Ben-Avie, & J. Ensign (Eds.), *How social and emotional development add up: Getting results in math and science education* (pp. 36–64). New York, NY: Teachers College Press.

Chapter 9
Changing STEM and Entrepreneurial Thinking Teaching Practices and Pedagogy Through a Professional Learning Program

Lihua Xu, Coral Campbell and Linda Hobbs

9.1 Introduction

There is a strong emphasis, both in Australia and internationally, on the importance of Science, Technology, Engineering, and Mathematics (STEM) education across all levels of schooling. In Australia, the push for schools to implement STEM education is reflected in government policy documents, such as the *National Innovation and Science Agenda* (Department of the Prime Minister and Cabinet, 2015), and the endorsement of the National STEM School Education Strategy 2016–2026 by all education ministers to focus on the development of mathematical, scientific, and technological literacy and twenty-first century skills (Education Council, 2015). The challenge for schools and those supporting teacher and school change is to translate a STEM policy agenda into valid and coherent curricula.

STEM competencies are regarded as essential not only for STEM occupations, but as the basis of a capable and informed citizenry (see Marginson, Tytler, Freeman, & Roberts, 2013). STEM offers a range of benefits for education and career opportunities of future generations (Timms, Moyle, Weldon, & Mitchell, 2018). On the one hand, quality STEM education allows for the exploration of STEM disciplines as interconnected bodies of knowledge and practices and a reconceptualization of knowledge and skills for the twenty-first century and beyond. On the other hand, STEM is believed to be a promising approach to draw more students into the future STEM workforce and offer enhanced career opportunities for these students.

Education in STEM begins in primary schools (Prinsley & Johnston, 2015). Yet, research shows the lack of confidence and competence of primary school teachers in teaching STEM-related areas such as science and mathematics (Marginson et al., 2013; Tytler, 2007). These have been regarded as one of the contributing factors to low student interest in STEM-related areas (Schreiner & Sjøberg, 2004), reduced

L. Xu (✉) · C. Campbell · L. Hobbs
Faculty of Arts and Education, Deakin University, Burwood, Australia
e-mail: lihua.xu@deakin.edu.au

© Springer Nature Singapore Pte Ltd. 2019
Y.-S. Hsu and Y.-F. Yeh (eds.), *Asia-Pacific STEM Teaching Practices*,
https://doi.org/10.1007/978-981-15-0768-7_9

student performance in international assessments (Marginson et al., 2013; Timms et al., 2018), and declining uptake of STEM beyond compulsory years (Kennedy, Lyons, & Quinn, 2014; Thomas, Muchatuta, & Wood, 2009; Tytler, 2007; Wienk, 2017).

Despite an increasing number of approaches emerging to upskill primary school teachers and build their capability for teaching STEM-related areas (Office of the Chief Scientist, 2016), the challenges surrounding implementation of STEM education in primary schools remain. The most critical challenge is how to meaningfully embed STEM-related knowledge, skills, and dispositions across all grade levels and curriculum areas through a whole-school approach. Increased quality and quantity of STEM implementation in primary schools requires not only the development of teachers to better understand and integrate content and pedagogical knowledge in STEM areas, but also their ability to lead changes in their schools in relation to teacher learning, curriculum, and instruction.

In this chapter, we report on school responses to a professional learning program designed to build teacher capacity for STEM teaching. Research questions for analysis are as follows:

- What aspects of the professional development program built teacher confidence and capacity for STEM teaching through inquiry-based approaches?
- How did the teachers and schools respond to the professional learning program?

9.2 STEM Practices in Primary Schools

Primary school teachers in Australia are generalists who are usually responsible for teaching all subject areas, including mathematics and science. For decades, research has shown that primary teachers lack confidence and competence in teaching these two subject areas and that this continues to be a problem in Australia and internationally (Marginson et al., 2013; Roth, 2014). Recent initiatives to address this situation include recommending mathematics and science subject specializations in preservice primary teacher education courses (Teacher Education Ministerial Advisory Group, 2014), improving primary teachers' science and mathematics subject matter knowledge, providing teachers with specific curriculum materials (Davis, Janssen, & van Driel, 2016; Hackling, Peers, & Prain, 2007), and improving teachers' knowledge of specific teaching approaches (Herbert, Xu, & Kelly, 2017). However, these initiatives typically focus on the *top-up* of teachers' content and pedagogical knowledge for teaching STEM subjects, rather than a more holistic approach that can be adapted by teachers to develop their own approach to teaching and using STEM practices.

STEM subjects have tended to be taught separately in primary schools due to existing curriculum structures. In the Australian curriculum (Australian Curriculum, Assessment, and Reporting Authority [ACARA], 2016), STEM is addressed

through the learning areas of Science, Technologies and Mathematics, and through general capabilities, particularly Numeracy, Information and Communication Technology (ICT) capability, and Critical and Creative Thinking. … Engineering is addressed in Design and Technologies through a dedicated content description at each band that focuses on engineering principles and systems. It is presented across the curriculum through Science, Digital Technologies and Mathematics. Engineering often provides a context for STEM learning. (p. 6)

Increasingly, there is pressure to teach science and mathematics through interdisciplinary or integrated approaches (English, 2016) with the aim to develop students' interdisciplinary understanding of STEM subjects, their STEM skills such as problem-solving and modeling (Ríordáin, Johnston, & Walshe, 2016), and to help students recognize the role of STEM in many aspects of their lives. Vasquez, Sneider, and Comer (2013) proposed a framework using four levels to present increasing degrees of integration for STEM areas: from disciplinary, to multidisciplinary, to interdisciplinary, to transdisciplinary. As the level of integration increases, there is more emphasis on the interconnections among different disciplines and the ability of students to apply knowledge and skills from each discipline to solve real-world problems. Bryan, Moore, Johnson, and Roehrig (2015) suggested that the integration of STEM needs to be specific and purposeful with the consideration of both *content* and *context*. They proposed three forms of STEM integration: (a) content integration in which multiple STEM objectives are included in the learning process, (b) integration of supporting content in which the learning objectives of one area is supported by another area, and (c) context integration in which the context of one area is used for achieving the learning of the objectives in another area.

Despite the increased number of approaches proposed in the literature for STEM integration in schools, research highlights some ongoing challenges, and complexities to navigate through the current STEM landscape and to provide effective approaches to STEM teaching practices (English, 2017). There are several significant issues to be considered, including how the implementation of STEM may interfere with the current school timetable and curriculum, how conversations between teachers from different disciplines can be encouraged, and how to meaningfully integrate STEM disciplines.

Professional learning programs that address some of these challenges have been developed and implemented across Australia (see Timms et al., 2018, for a list of government STEM policies and programs). Such professional learning programs tend to focus on activities, subject matter content knowledge, curriculum resources, and pedagogical strategies as approaches to teacher capacity building—the intention of which is to generate change in current practice and can include dispositions, knowledge, skills, attitudes, and motivation (Fullan, 2005). One example of such an initiative is the Primary Mathematics and Science Specialists (PMSS) program run by the Victorian Department of Education and Training. The focus of the PMSS program is on the education of primary teachers to become science or mathematics specialists, with the intention that these specialists would then lead capacity building for other teachers within their schools (Campbell & Chittleborough, 2014). Given the subject-specific focus of these initiatives, what is missing is attention to the broad structures (including frameworks and language) to allow for the organic development of a vision framework for guiding STEM practices in schools.

9.3　The STEM and Entrepreneurship in Primary Schools (SEPS) Program

STEM and Entrepreneurship in Primary Schools (SEPS) is a professional learning program designed to build teacher confidence and capacity for STEM teaching through inquiry-based approaches. SEPS was funded by the Australian Department of Industry, Innovation, and Science (https://www.business.gov.au/). The SEPS program aimed to:

- build and improve teacher capability and innovation in the teaching of STEM and entrepreneurship programs in primary schools in the Geelong region;
- raise awareness among primary school students of the value of STEM and entrepreneurship; and
- increase participation among primary school students in STEM activities and engagement in entrepreneurial challenges.

In 2018, 11 primary schools from the Geelong region of the state of Victoria participated in three key events run by the Deakin team: a professional learning (PL) workshop, a student maker faire, and a STEM education conference. Two teachers from each school were committed to the PL programs for the entire year. At the start of the program, teachers participated in a 2-day PL workshop at Deakin University. The foci of the two intensive days were to (a) introduce conceptual frameworks for guiding STEM curriculum planning and (b) build teachers' knowledge of the variety of STEM practices and pedagogies, enabling them to make informed and effective decisions relating to STEM education in their own school context.

After the 2-day workshop, the teachers planned and implemented a STEM program in their schools. Some of the students' work generated from the STEM programs was displayed at the Maker Faire hosted at the university. At the Maker Faire, 180 students participated in a range of activities, including sessions led by students, teachers, and invited scientists. The last event of the SEPS program was a 3-day national STEM education conference at the university (https://www.deakin.edu.au/stem-education-conference-2018) at which the participating teachers showcased their innovations and key learnings to a delegation of teachers, educators, and educational researchers.

Prior to joining this program, these schools were at different points in relation to their development and implementation of STEM programs. While a few schools were at the very beginning of establishing a STEM program and resources, others had been implementing it for some time, with one school offering a STEM program for 4 years. However, these existing STEM programs tended to be run separately either in the form of extracurricular activities for a small group of students, embedded in one of the curriculum areas (e.g., digital technology), or as a separate program run by a specialist teacher. No school had established a whole-school approach in which STEM was embedded across multiple curriculum areas and year levels. In order to cater for the different needs of the participating schools, the SEPS program was designed to provide teachers with conceptual frameworks for the purpose of guiding their thinking in relation to both planning and implementing programs on STEM and entrepreneurship in their schools. Schools were expected to decide their own development goals. As the program progressed, it was anticipated that the teachers would focus not only on their own practices but also on work with other teachers to lead sustainable STEM innovation across their school.

The 2-day workshop incorporated two frameworks: the SEPS Guiding Framework and the STEM Entrepreneurial Thinking (STEM-ET) Vision Framework. These frameworks provided a comprehensive, multifaceted, coherent approach to addressing the subtle and complex challenge of preparing twenty-first century citizens within the constraints of a traditional school system and curriculum. The frameworks give teachers a common language that could be operationalized to support teacher and school change.

9.3.1 SEPS Guiding Framework

The SEPS Guiding Framework (Fig. 9.1) was developed by the Deakin University team for this project and incorporates earlier STEM research into innovation through technology (Albion, Campbell, & Jobling, 2018) and STEM practices (Hobbs, Cripps Clark, & Plant, 2018). The framework comprises four interconnected components to support schools and teachers to frame and plan for curriculum and programs with a strong focus on STEM and Entrepreneurial Thinking (ET).

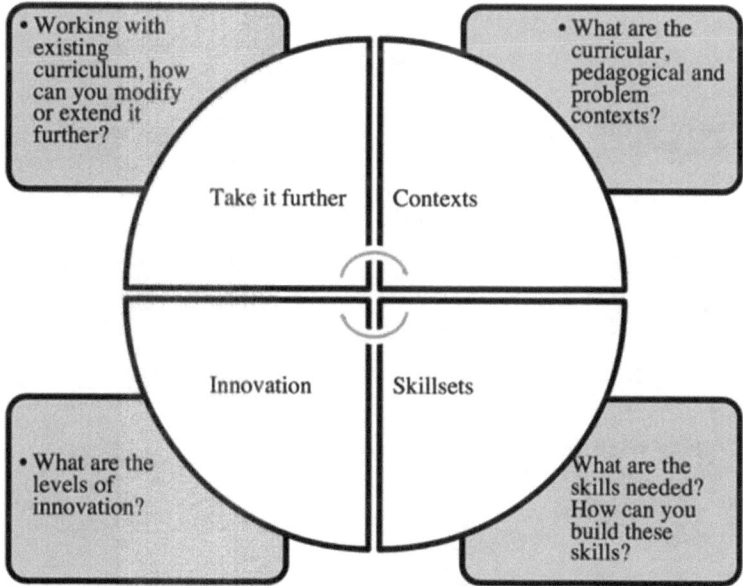

Fig. 9.1 SEPS guiding framework

The four components of the SEPS Guiding Framework—contexts, skillsets, innovation, and take it further—allow schools to work through the process of identifying a problem area within their curricula and school context for which a STEM program incorporating entrepreneurship (innovation) can be developed and implemented. It was expected that the SEPS Guiding Framework would be used in two ways: as a framework to guide the development of STEM-ET programs for the whole school and as a conceptual tool for teachers to develop their own approach to STEM-ET teaching and learning.

9.3.2 A STEM-ET Vision Framework for Teachers and Schools

The STEM-ET Vision Framework (Fig. 9.2) is an extension of the STEM Vision Framework developed by Hobbs et al. (2018). By incorporating entrepreneurial thinking, the STEM-ET Vision Framework enables educators to engage with the STEM agenda through a targeted and deliberate framing of STEM linking with ET;

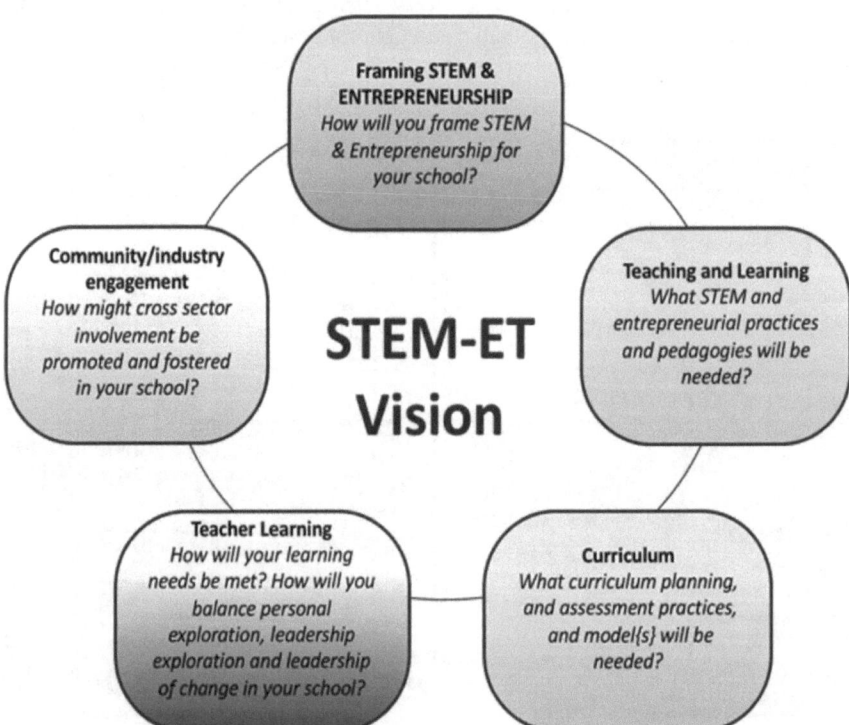

Fig. 9.2 STEM-ET vision framework (adapted from the Successful Students—STEM Program [Hobbs et al. 2018])

the STEM practices that can inform teaching, learning, and curriculum; how to build teacher capacity and lead change in schools; and considerations to community and industry links as meaningful contexts for learning.

The STEM-ET Vision Framework can be used in the following ways:

- As a vision for an approach to STEM-ET education guiding a professional development project.
- For a school to develop a vision for STEM-ET that can function at multiple levels within the school.
- For a teacher to develop their own approach to STEM-ET education, including a professional development plan.

9.4 Overview of Approaches to STEM Education by the SEPS Schools

The Deakin University team evaluated each of the three SEPS program events (i.e., PL workshop, maker faire, conference) to ascertain changing attitudes, perceptions, and knowledge of students and teachers in relation to STEM and ET. In addition to the evaluation of each event using questionnaires, additional data were gathered to track the changes in teacher capacity and student engagement in STEM and entrepreneurship activities and practices. The research methods include document analysis and semi structured interviews with teachers and principals in case study schools. Case studies were undertaken in six SEPS schools to enable documentation of teacher, curriculum, and school change as a result of teachers' SEPS participation and to establish the challenges and success criteria influencing the successful and sustainable implementation of STEM/entrepreneurship initiatives. All schools were invited to be part of the case studies; however, for various reasons, five opted out.

For the purpose of identifying how the program influenced teacher and school change, data from interviews with SEPS teachers and sometimes principals from the six case study schools were analyzed to identify to what extent and how STEM (or for some schools STEAM with A as the Arts) was being implemented prior to the SEPS program, what initiatives or actions that the SEPS teachers undertook during the program, and their future plans. Ascertaining future plans enabled us to gain insights into changes that might occur beyond the formal SEPS program. These findings are summarized in Table 9.1. Following the table is a summary of some of the variations and commonalities across the six case study schools.

On entry to the program, teachers' STEM teaching capability and school STEM programs were at different stages of maturity. These different stages of maturity are discussed below in terms of the degree of progress made in the schools.

School M was at entry level as it had no explicit STEM programs. The motivation to embrace STEM came mainly from the school leadership rather than the teachers participating in the SEPS program; this, combined with a lack of STEM

Table 9.1 Summary of Case Study School STEM Programs before, during, and after SEPS

School	Pre-SEPS STEM initiatives and directions	Implemented STEM initiatives (terms 3 and 4 of 2018)	Future plans
A	1 h stand-alone lesson per week for all grade levels. Rotation of activities focusing on digital learning. No integration across subjects	Grades 3–4 (Grade 3/4): STEM rotation of theme-based activities that are based on challenge, focused on design thinking, incorporating technology, and integrated	Whole school integrating STEM into everything rather than just stand-alone sessions
B	No STEM program. Some inquiry units had science focus. Science taught by a science specialist. Digital technology taught by a specialist. Some individual teachers dabbled with STEM ideas and digital technologies	Grades 3/4, 5/6: War on Waste project where students worked in one of 10 disciplinary groups (e.g., mathematics, science, reading, physical education, art) to solve a waste-related problem. Digital technology training for staff through professional development	Grades 2–6: Class-based projects using role-based groups to solve complex problems that were determined by the teachers and students
C	Whole-school innovate project where teachers worked in teams to investigate new ideas, including STEAM and sustainability teams. Development of tinker boxes for classroom teachers to implement	Whole school: Continued implementation of the tinker boxes, with Grades Preparatory (Prep)–2 box set and Grades 3–6 box set. Each box includes construction, coding, and design activities. Design as individual (station), small group, or whole-class activities. Grades Prep, 5/6: Inquiry- and problem-based activities incorporating the design process as part of the themed units' implementation by specific teachers. Grade 5/6: entrepreneurial activities (e.g., shark tank)	None identified

(continued)

Table 9.1 (continued)

School	Pre-SEPS STEM initiatives and directions	Implemented STEM initiatives (terms 3 and 4 of 2018)	Future plans
F	Science specialist program 1 h for two terms for Grades 3–6. STEAM class being introduced in 2019 in response to new school direction and new building with dedicated space	Grades 3–4, 5–6: specialist-led STEAM classes where students complete engineering-based activities focusing on key areas of the curriculum, such as sustainability, forces (through hydraulics, and pneumatics construction). Some use of mathematics, science, and technology on a needs basis. Partnerships with other schools and industry groups	Incorporate Grade 1/2 into the rotation to allow more students to be involved. Work with classroom teachers to embed elements of the programs where there is a natural fit to their curriculum
M	Inquiry topics, stand-alone. Not labeled as STEM-related	STEM lesson 1 h per week. STEM is a focus of planning and delivery by interconnecting the STEM lessons with the current classroom inquiry topic	None identified
R	STEM specialist program 1 h per week; focus was mainly on science and some incorporation of technology	Problem based where problem is posed by teacher and students go through the problem-solving and design processes to solve those problems (e.g., How do we use government money for our new oval?). Link STEM activities with what the classroom teachers are doing	Topic-based units to cover a number of STEM areas

teaching experience, meant that progress during the SEPS program focused mainly on modifying existing programs.

School B was also at an entry level with no preexisting STEM program; however, motivation to be involved in SEPS came from the teachers as a result of a school community decision to embrace STEM. Strong teacher motivation and buy-in from other teachers meant that substantial progress was made toward implementing a STEM unit as a result of the contacts and ideas from the SEPS PL and from other professional development and conferences in 2018.

School A had made some progress toward implementing STEM as a set of rotation activities that had a strong focus on digital learning. These activities were redeveloped in 2018 in order to make them more STEM-like, which meant incorporating more challenges and design elements and moving toward interdisciplinarity.

School C had made progress with STEAM implementation in 2017 through a program of teacher research, which resulted in the introduction of tinker boxes to all classrooms. This was continued in 2018 with an expectation that all teachers spend 50 min every fortnight on activities. In addition, the SEPS teachers made adjustments to their teaching by introducing project-based learning at Prep and Grade 5/6 as well as an entrepreneurial activity in Grade 5/6.

School R had made some progress toward having a specialist deliver STEM with a strong science focus. However, the specialist teacher changed part-way through 2018, and the new teacher (the remaining SEPS teacher) in this role was at a preliminary stage in conceptualizing how best to frame and structure the STEM program. Some progress was made in conceptualizing how to integrate the specialist programs offered by the school as well as align the specialist STEM program with the classroom inquiry focus. By the end of the year, the SEPS teacher had moved toward using a problem-based approach during STEM lessons. For example, an inquiry unit called Beneath My Feet being taught by Grades 3–6 teachers were supported by the STEM specialist through a problem-based task focused on designing a new school oval; it included questions relating to processes of preparing the ground, formal and estimated dimensions and quantities, and financial literacy. A Paddock to Plate inquiry unit in Grades 1–2 enabled a focus on lifecycles and hatching eggs in the STEM classes.

School F had made substantial progress in their implementation of STEAM prior to SEPS, with a new STEAM specialist class in 2018 building on a mature science specialist program in previous years. There was substantial commitment from the school leadership to developing a strong STEAM program that tied into their school values of *innovative, collaborative,* and *inspiring*; it was facilitated by a newly built dedicated STEAM teaching space. STEM units were implemented in 2018 focusing on design-based learning. For example, a hydraulics and pneumatics unit could be linked to the science and mathematics curriculum where relevant, such as push and pull in Grades 1/2 and force and resistance in Grades 3/4.

These variations in entry points and progress highlight the need to be flexible when planning professional development so that there are scope and potential for all participants to attain success. Certainly during the 8 months of the SEPS program, there was evidence of the SEPS teachers being involved in curriculum design and pedagogy change as well as leading change in their schools. Some teachers referred to running professional development for the whole teaching staff and specific teaching teams to encourage new activities or pedagogies, build teacher ability, and conceptualize or reconceptualize STEM or STEAM for their school. School C had done some of this work in 2017, while Schools B, F, and A mentioned specific professional development delivered to other teachers in 2018. Across the six schools, there was evidence of existing activities being revamped or redeveloped (Schools A and B), introduction of new pedagogies or activities such as problem-based learning

(Schools R and C) or project-based learning (School B), entrepreneurial thinking either explicitly or implicitly through a focus on innovation activities (Schools B and C), and design-based activities using a design cycle (all schools).

A case study follows that illustrates how the conceptual frameworks can be implemented. School B was selected because the teachers and school were relatively new to implementing comprehensive STEM programs and there was clear evidence of all parts of the SEPS Guiding Framework in their description of activities in 2018.

9.5   Case Study of a SEPS School: School B

School B is a small primary school in a Victorian regional city. The school has a student population of 191 students with about 40% from diverse backgrounds. The school's Index of Community Socio-Educational Advantage indicates a disadvantaged school population; as such, there is need for greater support for this student body and community. Many of the school's current programs focus on well-being and student's emotional health and well-being and resilience development (https://www. myschool.edu.au/). The broader school community has tended not to be strongly involved in school programs (Interview, Teacher T, Nov. 2018).

According to the two SEPS teachers (both teaching Grade 3/4), prior to 2018, there was no official STEM program at the school: "STEM didn't exist" (Interview, Teacher S, Nov. 2018). They had been running some science and digital technology programs, and there had been a move t o incorporate a "range of *emerging* STEM practices that relate directly to creating problem-solving, innovative young children with a sense of entrepreneurship" (Expression of Interest, Nov. 2017). In the last 2 years, the school has had a digital technologies teacher who teaches coding; the school has a one-to-one iPad program (i.e., all students have their own device) for Grades 3–6. In addition, a specialist science teacher has been teaching separate science classes through a 2-year curriculum cycle that the classroom teachers supplemented. Other activities included participation in a nation-wide Robocup competition, attempts by some teachers to introduce STEM in their classrooms, and involvement in a science-focused leadership program with the local secondary school for the Grade 6 children. The two SEPS teachers had been developing units that, while focused on the science topic, were becoming more integrated and more STEM-like.

In response to the SEPS PL, the two SEPS teachers took responsibility for reviewing the school's current program offerings. They purchased a range of technological equipment, including programmable items such as Spheros™ (https://www.sphero. com/education/), Makey Makey® (https://makeymakey.com/), and virtual reality headsets. Most significantly, they redeveloped an old unit that no longer worked to make it focused on STEM for Grades 3–6. After attending the SEPS PL and talking with other teachers in the network, the two SEPS teachers decided to introduce a problem-based approach. Their first attempt was to create a problem for the students to solve: turning unused cricket nets into a chicken coop. They realized that this was too big a project to start with as the children could not cope with the open problem-solving and needed more guidance. They then worked with the other Grade 3/4 teachers to take a project-based approach and used a STEM-based theme focused on a common global issue or problem. They selected the theme War on Waste that was based on a recent Australian Broadcasting Corporation television series of the same name.

The unit used the existing group working structure of the old unit where students worked in discipline-based groups. After watching some of the War on Waste television episodes, the students worked in ten groups, each exploring the theme through the lens of a different discipline (e.g., mathematics, writing, physical education, science, etc.). Students were assisted by their teachers to develop a focus and integrate information across the groups where appropriate. For example, one group investigated the science of composting and the mathematics group collected data on waste composition. These data were then given to the composting group to support their analysis and the writing group to prepare a persuasive text.

Through collaboration between groups, the teachers wanted the students to realize that the problem was multifaceted and, therefore, required a wide range of thought processes to arrive at multiple solutions. They identified problem-solving, critical thinking, and critical analysis as the skills they hoped students would gain through their project work. The teachers also raised students' awareness of the problem in different contexts: waste management locally at their school, waste in the local community, and effects of waste globally. In terms of entrepreneurial approaches, the STEM program was focused on engagement, but the teachers recognized the entrepreneurial achievements of their students in some of the work they completed in class, indicating that they "encourage (sic) entrepreneurial thinking … when it is presented" (Interview, Teacher T, Nov. 2018).

An aim of the 2018 STEM program was to enhance student autonomy, but the SEPS teachers recognized that with some students' limited problem-solving skills more teacher control was needed initially. Furthermore, they found that a project-based approach was more suited to the students' current learning needs. Because this was a pilot and teachers were still learning how to support this type of learning, rather than using this as a formal assessment opportunity, the teachers focused on engaging the students in authentic learning: "at this point in time, [it's] not assessed" (Interview, Teacher T, Nov. 2018). However, the SEPS teachers referred to changes in behavior relating to recycling and in attitude toward waste and the environment and regarded this as a worthwhile outcome of the unit.

An important outcome of attending the SEPS PL was that the SEPS teachers felt enabled to lead the direction of STEM in their school. During the year they provided professional development for staff, particularly in the areas of digital technologies, "to build staff competency" (Interview, Teacher S, Nov. 2018) so that they were familiar with the technology and could seek opportunities to integrate it into their regular classroom teaching. Since this in school professional development, some teachers in the Grade 1/2 team used some of the digital tools in their teaching.

The SEPS teachers reported that the opportunity to network with other SEPS teachers was particularly valuable in seeing several models of STEM implementation in practice. Speaking with the other teachers enabled them to choose aspects of STEM practices that were most appropriate to their own schools' needs.

> Obviously, there is an overarching network thing there. I suppose the schools we have taken a bit from, probably School G [not a case study school], School F and then from the [STEM] conference, also School A and the way they have developed their inquiry. (Interview, Teacher T, Nov. 2018)

For the future, the SEPS teachers intend to work with the other Grades 2–6 teachers to redesign the unit so that the students explore a theme for the entire year. The theme will be determined by the classroom teacher; the intentions are that more digital technology will be used where appropriate, that students will build on their problem-solving skills, and that teachers will guide greater collaboration across the student groups. The year will culminate in a whole-school exposition of students' achievements involving the broader communities of parents, other schools, and local industry.

The case study of School B provides a good example of progress made by a school with limited existing STEM programs toward a purposefully built STEM program to enable authentic learning experiences for its students. The SEPS teachers indicated that things are moving slowly and steadily but much depends on individual teacher buy-in. Overall, the school presents a picture of engagement in STEM across most levels with a number of significant emerging programs in development. Leadership support is available, and the teachers' commitment to bring about change in STEM was evident. The role of STEM leader was introduced in 2018, also demonstrating the school's commitment to implementing a quality and comprehensive STEM program. The gradual release model employed by the two SEPS teachers in implementing the STEM program attracted some initial buy-in from students and some teachers. This was perceived as a stepping-stone toward building teacher capacity in the school in order to develop a whole-school approach for STEM.

9.6 Discussion

Description of the six case study schools who participated in the SEPS program demonstrated that a range of models were employed by the schools to implement STEM, including

- Classroom teachers or whole departments (e.g., all Grades 3 and 4 classes) implement a STEM hour (Schools B and C).
- Classroom teachers deliver STEM as the focus of the inquiry unit (Schools C, B, and M).
- STEM/science/technology specialist classes run separately to classroom teaching (Schools A, B, C, F, M, and R).
- STEM specialist classes to enhance regular classroom programs (Schools M and R).
- Specialist teacher classes work together (School R)

- Individual classroom teachers incorporate STEM pedagogies where they can (School C).

There was concerted effort by all the case study schools to consider moving away from disconnected teaching of separate STEM areas (e.g., focused discretely on science or digital technology) toward a more coherent, cross-disciplinary student experience. For the schools that started with a specialist model, attempts were made to connect the specialist program with regular classroom teaching so as to deliver a more coherent STEM curriculum across the school. For the schools in which STEM was built into the timetables of regular classroom teaching (e.g., inquiry unit), efforts were made to embed STEM into other curriculum areas.

The SEPS program was designed to allow flexibility for schools in what they decided to work on, improve, or develop while undertaking the PL. We offered frameworks (i.e., Guiding Framework and Vision Framework) that schools could select from and be informed by rather than promoting particular pedagogies or approaches (e.g., digital technologies, design-based thinking). It should be noted that PL in a range of pedagogies and approaches was certainly part of the 2-day workshop. Similar to the outcomes of the professional development reported in Hobbs et al. (2018), which was based on a similar flexible approach to teacher professional development, the variation in models of approaching STEM by the six case study schools illustrate how the two conceptual frameworks could support the needs of each school in relation to teacher PL, STEM curriculum development, and STEM teaching and learning, regardless of their starting point.

The four components of the Guiding Framework (Fig. 9.1), in particular, allowed each school to develop its own goals for STEM improvement based on where they were on entry to the program. The contexts component of the Guiding Framework enabled teachers to decide what is relevant and meaningful for the students to work on as authentic or real-life problems. This component arose out of advice received from schools while planning the SEPS program that primary teachers need support to develop meaningful links with community and industry. The framework recognizes that schools tap into immediate, school, local, and global contexts while delivering their curricula. School B demonstrated this well through their War on Waste theme, with school, city, and global waste-related issues being explored by the students. Dealing with authentic problems and contextualizing the curriculum are critical to making STEM relevant for students' current and future lives (Darby-Hobbs, 2013; Tytler, 2007).

The focus on skill sets provided the teachers with a common language to articulate their knowledge of STEM and STEM education in meaningful ways to others participating in SEPS as presentations at the STEM education conference and maker faire, to other teachers in their schools by running professional development through school-based meetings, and for focusing student learning and linking their programs to the curriculum. Some of the key skills highlighted in the interviews include problem-solving, critical thinking, and creativity.

The introduction of entrepreneurial thinking to the program added a new dimension to professional learning programs focusing on STEM. Entrepreneurship was featured in two components of the Guiding Framework: as one of the skills in the skill set and innovation components. A few schools started to build ET as part of their STEM programs. For example, the Shark Tank entrepreneurship program in School C was modeled off a popular TV program of the same name where the contestants propose new inventions to meet a designated need. Other schools, such as School F, recognized that ET is valuable and have plans to incorporate it explicitly in the future where relevant. School B found that where students had autonomy and choice in how they responded to the problem of waste, some engaged with ET. For example, a 10-year-old girl devised the Straw No More Campaign banning straws in pubs, clubs, and school canteens; another Grade 5/6 science group created a retirement plan based on the retrievable gold in unused mobile phones.

The last component—Take it further—allowed schools to adapt, modify, or extend existing curricula and programs to optimize student learning. Such attempts were evident across all six case study schools, which are illustrated by School B with its adaption of an existing program to a new format and adoption of the new War on Waste theme. This component of the Guiding Framework recognizes that schools have often made some moves to developing innovative STEM programs. The teachers were in no way compelled to develop new activities as part of the PL program but were encouraged to think about how they could adapt existing activities or programs.

9.7 Conclusion

In this chapter, we described a professional learning program specifically designed to build primary teachers' ability to design and deliver STEM programs in their schools. Despite the relative short length of the program (i.e., 8 months), the changes occurring in schools are apparent. We argue that the SEPS Guiding Framework and the STEM-ET Vision Framework can play an important role in facilitating change in schools. These two frameworks provided ways for schools to decide how to respond to the STEM agenda and how this can be enacted at various levels within the school—whether by new programs, refocusing existing programs, modifications to teacher pedagogies, new structures within the school, or new STEM leaders. These two conceptual frameworks can be used to support the development of a STEM vision for the whole school and the design of individual STEM programs that are grounded in school context; they can also be a tool for facilitating teacher professional learning within the school.

In a recent review of STEM literature and policy, Timms et al. (2018) recommended three strategies for rethinking the STEM curriculum:

- Work from an agreed definition of STEM curriculum.
- Shift to an emphasis on STEM practices.
- Move toward an integrated STEM curriculum. (p. 25)

Our project demonstrates that those programs that worked in schools are home grown and well-grounded in the context of the school, taking into consideration the capacities of the teachers and students, available resources, and constraints. The diversity in both conceptualizing and working with STEM provides teachers with opportunities to innovate in their responses to the STEM agenda and to learn from each other. A one-size-fits-all approach is less likely to meet local issues and enable real changes in practice. Future professional learning programs need to consider multifaceted approaches to enable teacher learning and facilitate sustainable school changes in the era of STEM.

Acknowledgements We acknowledge the Department of Industry, Innovation, and Science, Australia, for funding this project as well as Skilling the Bay and Upstart Challenge for their support of this project. We would also like to thank the participating teachers who are open to new ways of working and embracing challenges brought by the STEM agenda.

References

Australian Curriculum, Assessment, and Reporting Authority. (2016). *ACARA STEM connections project report*. Canberra, ACT: Author. Retrieved from, https://www.australiancurriculum.edu.au/media/3220/stem-connections-report.pdf.

Albion, P., Campbell, C., & Jobling, W. (2018). *Technologies education for the primary years*. South Melbourne, Australia: Cengage.

Bryan, L. A., Moore, T. J., Johnson, C. C., & Roehrig, G. H. (2015). Integrated STEM education. In C. C. Johnson, E. E. Peters-Burton, & T. J. Moore (Eds.), *STEM road map: A framework for integrated STEM education* (pp. 23–37). New York, NY: Routledge.

Campbell, C., & Chittleborough, G. (2014). The "new" science specialists: Promoting and improving the teaching of science in primary schools. *Teaching Science: Journal of the Australian Science Teachers Association, 60*(1), 19–29.

Darby-Hobbs, L. (2013). Responding to a relevance imperative in school science and mathematics: Humanising the curriculum through story. *Research in Science Education, 43*(1), 77–97.

Davis, E. A., Janssen, F. J. J. M., & van Driel, J. H. (2016). Teachers and science curriculum materials: Where we are and where we need to go. *Studies in Science Education, 52*(2), 127–160.

Department of the Prime Minister and Cabinet, Commonwealth of Australia. (2015). *National innovation and science agenda*. Canberra, ACT: Author. Retrieved from, https://www.industry.gov.au/data-and-publications/national-innovation-and-science-agenda-report.

Education Council. (2015). *National STEM school education strategy, 2016–2026: A comprehensive plan for science, technology, engineering, and mathematics education in Australia*. Carlson South, Australia: Author http://www.educationcouncil.edu.au/site/DefaultSite/filesystem/documents/National%20STEM%20School%20Education%20Strategy.pdf.

English, L. D. (2016). STEM education K–12: Perspectives on integration. *International Journal of STEM Education, 3*(3), 1–8.

English, L. D. (2017). Advancing elementary and middle school STEM education. *International Journal of Science and Mathematics Education, 15*(1), 5–24.

Fullan, M. (2005). *Leadership & sustainability: System thinkers in action*. Thousand Oaks, CA: Corwin Press.

Hackling, M., Peers, S., & Prain, V. (2007). Primary connections: Reforming science teaching in Australian primary schools. *Teaching Science, 53*(3), 12–16.

Herbert, S., Xu, L., & Kelly, L. (2017). The changing roles of science specialists during a capacity building program for primary school science. *Australian Journal of Teacher Education, 42*(3), 1–21. https://doi.org/10.14221/ajte.2017v42n3.1.

Hobbs, L., Cripps Clark, J., & Plant, B. (2018). Negotiating partnerships in a STEM teacher professional development program: Applying the STEPS interpretive framework. In L. Hobbs, C. Campbell, & M. Jones (Eds.), *School-based partnerships in teacher education: A research informed model for universities, schools and beyond* (pp. 231–246). Dordrecht, The Netherlands: Springer.

Kennedy, J., Lyons, T., & Quinn, F. (2014). The continuing decline of science and mathematics enrolments in Australian high schools. *Teaching Science, 60*(2), 34–46.

Marginson, S., Tytler, R., Freeman, B., & Roberts, K. (2013). *STEM: Country comparisons. International comparisons of science, technology, engineering and mathematics (STEM) education.* Report for the Australian Council of Learned Academies. Melbourne, Australia: ACOLA. Retrieved from, https://acola.org.au/wp/PDF/SAF02Consultants/SAF02_STEM_%20FINAL.pdf.

Office of the Chief Scientist. (2016). *STEM programme index 2016.* Canberra, ACT: Australian Government.

Prinsley, R., & Johnston, E. (2015). *Transforming STEM teaching in Australian primary schools: Everybody's business.* Canberra, ACT: Australian Government.

Ríordáin, M. N., Johnston, J., & Walshe, G. (2016). Making mathematics and science integration happen: Key aspects of practice. *International Journal of Mathematical Education in Science and Technology, 47*(2), 233–255.

Roth, K. J. (2014). Elementary science teaching. In N. G. Lederman & S. K. Abell (Eds.), *Handbook of research on science education* (2nd ed., pp. 361–394). London, England: Taylor & Francis.

Schreiner, C., & Sjøberg, S. (2004). *Sowing the seeds of ROSE: Background, rationale, questionnaire development and data collection for ROSE - a comparative study of students' view of science and science education.* Oslo, Norway: Department of Teacher Education & School Development, University of Oslo.

Thomas, J., Muchatuta, M., & Wood, L. (2009). Mathematical sciences in Australia. *International Journal of Mathematical Education in Science and Technology, 40*(1), 17–26.

Teacher Education Ministerial Advisory Group. (2014). *Action now: Classroom ready teachers.* Canberra, ACT: Author. Retrieved from https://www.aitsl.edu.au/tools-resources/resource/action-now-classroom-ready-teachers.

Timms, M., Moyle, K., Weldon, P., & Mitchell, P. (2018). *Challenges in STEM learning in Australian schools: Literature and policy review.* Melbourne, Australia: Australian Council for Educational Research. Retrieved from https://research.acer.edu.au/cgi/viewcontent.cgi?article=1028&context=policy_analysis_misc.

Tytler, R. (2007). *Re-imagining science education: Engaging students in science for Australia's future.* Victoria, Australia: Australian Council for Educational Research.

Vasquez, J. A., Sneider, C., & Comer, M. (2013). *STEM lesson essentials, Grades 3-8: Integrating science, technology, engineering, and mathematics.* Portsmouth, NH: Heinemann.

Wienk, M. (2017). *Discipline profile of the mathematical sciences 2017.* Melbourne, Australia: Australian Mathematical Sciences Institute. Retrieved from https://amsi.org.au/publications/discipline-profile-mathematical-sciences-2017/.

Chapter 10
Potential and Challenges in Integrating Science and Mathematics in the Classroom Through Real-World Problems: A Case of Implementing an Interdisciplinary Approach to STEM

Wanty Widjaja, Peter Hubber and George Aranda

10.1 Introduction

There has been a strong push to advocate science, technology, engineering, and mathematics (STEM) integration in education (English, 2016; California Department of Education, 2014; Marginson, Tytler, Freeman, & Roberts, 2013), yet the literature has been inconclusive as to what effective STEM integration entails. English (2016) called for systematic research into the effectiveness of integrated STEM education to develop students' knowledge of content in the respective disciplines.

Among the issues of concern are different interpretations of STEM integration, the nature and scope of such integration, and lack of balanced and transparent content representations in STEM (Bryan, Moore, Johnson, & Roehrig, 2015; English, 2016; English & King, 2015; Honey, Pearson, & Schweingruber, 2014; Marginson et al., 2013). Furthermore, issues related to the lack of agreement around the pedagogical approach to integrate the different disciplines remain unresolved (Leung, 2018).

Effective STEM integration entails explicit understanding of what STEM integration means and adequate knowledge of multidisciplinary content (Moore et al., 2014; Stinson, Harkness, Meyer, & Stallworth, 2009; Williams et al., 2016). One of the key barriers for integration documented in the literature is the different pedagogical traditions in science and mathematics (Nadelson & Seifert, 2017; Schoenfeld 2004; Tytler, 2016). Some researchers (e.g., Funner & Kumar, 2007) found that a lack of instructional resources, support materials, and pedagogical guidance for inquiry hinders teachers in linking and integrating science and mathematics in their classroom. Another key barrier to STEM integration is the pervading system of disciplinary silos in the school curriculum that is reflected in the teaching timetables (Gardner & Tillotson, 2018) as well as the boundaries between disciplines (Hobbs, 2012; Williams

W. Widjaja (✉) · P. Hubber · G. Aranda
Faculty of Arts and Education, School of Education, Deakin University, Burwood, Australia
e-mail: w.widjaja@deakin.edu.au

© Springer Nature Singapore Pte Ltd. 2019
Y.-S. Hsu and Y.-F. Yeh (eds.), *Asia-Pacific STEM Teaching Practices*,
https://doi.org/10.1007/978-981-15-0768-7_10

et al., 2016). Nadelson and Seifert (2017) argued that schools need to restructure their curriculum and alter the timing of instruction in order to create a school culture and environment that supports an integrated STEM approach successfully.

There are multiple approaches to designing an integrated STEM curriculum, which traverse along a continuum from a single discipline to a transdiscipline perspective (Vasquez, Sneider, & Comer, 2013; Williams et al., 2016). Leung (2018) proposed the use of an inquiry-based modeling pedagogical cycle as a hybrid pedagogy to cross between scientific investigation and mathematical modeling. He underscored the importance of establishing students' habits of mind (e.g., searching for uncertainty, recognizing ambiguity, and learning from failure) as key features of the inquiry modeling process. There is a need to understand constraints on innovations toward STEM integration at the system, school, and teacher levels.

In this chapter, we explore and discuss possibilities and challenges in integrating science and mathematics through real-world problems in the classroom, using a case study of two teachers from one secondary school that was implementing an interdisciplinary approach in STEM. In particular, this chapter addresses the research question: To what extent does the use of real-world problems support student engagement in interdisciplinary learning of science and mathematics?

10.2 Literature Review

10.2.1 STEM Integration Models

Empirical research on STEM integration remains elusive (Honey et al., 2014). Better delineation of what constitutes productive STEM integration is needed to enable theorization and conceptualization. There are multiple approaches to framing and designing integrated STEM curriculum (Bryan et al., 2015; English, 2016; Vasquez et al., 2013; Williams et al., 2016). Vasquez et al. (2013) proposed a model of integration that traverses along a continuum from disciplinary to multidisciplinary, interdisciplinary, and transdisciplinary.

The use of inquiry-based pedagogy as a pedagogical approach that supports integration across different disciplines in STEM has been documented in the literature (English, 2016; Lehrer & Schauble, 2000; Tytler, 2016). Real-world problems, modeling, and representation construction are central features to the interdisciplinary approach (English, 2009; English & King, 2015; Ferri & Mousoulides, 2017; Lehrer & Schauble, 2000; Tytler, 2016). Lehrer, Schauble, and Lucas (2008) asserted that building and refining models are central inquiry practices that characterize scientists' work. Studies conducted in the field of mathematics education (Blum & Niss, 1991; English, 2009, 2016) argue that mathematical modeling offers a vehicle for teachers to engage students in exploring authentic problems involving complex systems within an interdisciplinary context. This process is key in generating new knowledge and understanding in both mathematics and science. Furthermore, Bryan et al. (2015)

argued that processes and practices such as science inquiry, engineering design, and mathematical thinking and reasoning are essential in developing twenty-first century skills and should be an integral part of the STEM integration approach.

10.2.2 STEM Integration Issues

The central argument for adopting an integrative approach in teaching science and mathematics is driven by the intention to enable students to become aware of the links between mathematics and science, and engage them in meaningful and deeper learning (Ríordáin, Johnston, & Walshe, 2016; Treacy & O'Donoghue, 2014; Tytler, 2016). Corlu, Capraro, and Capraro (2014) underscored the importance of maintaining each STEM discipline's unique characteristics, depth, and rigor in the process of integration. English (2016) argued for a more transparent and balanced approach on each discipline in STEM integration and reiterated a similarly important point raised by Shaughnessy (2013) and Moore et al. (2014) about the need to lift the profile of mathematics and engineering in STEM integration.

Some processes and practices common to various disciplines can serve as an integrative theme in STEM programs. For example, problem-solving and modeling are very much part of mathematics and science. English (2009) advocated the use of modeling in both mathematics and science to start in primary school and "not be confined to the secondary school years and beyond" (p. 170).

Representations that express or symbolize an idea or relationship are important within each STEM discipline in the way knowledge is constructed and learned. Dreher, Kuntze, and Lerman (2016) pointed out that representations and their connections play a key role for experts in the construction of mathematical knowledge and for learners to build conceptual knowledge in the mathematics classroom. Johri, Roth, and Olds (2013) expressed similar views for the discipline of engineering, as did Latour (1999) for science. When considering the demands on the learner in an integrated STEM environment, Honey et al. (2014) argued that "Students need to be competent with discipline-specific representations and be able to translate between discipline-specific representations thereby exhibiting what some scholars refer to as 'representational fluency'." (p. 71).

Bryan et al. (2015) and English (2009) highlighted the importance of designing problems that are integrated into a classroom's particular learning theme rather than adding to an already crowded curriculum. Pearson (2017), reflecting on research since his work on the Honey et al. (2014) report, recommended that in integrated STEM contexts students should be supported in building knowledge and skills within and across disciplines and in deepening their knowledge and skills in individual disciplines. Finally, it was recognized that more integration is not necessarily better and that educators should be measured in the degree of integration according to their learning goals and their students' needs.

Researchers have identified different structures and pedagogical traditions in mathematics and science as factors that make crossing the boundaries between these disciplines difficult (Hobbs, 2012; Tytler, 2016; Williams et al., 2016). Other barriers to integration (Hobbs, 2012; Treacy & O'Donoghue, 2014; Tytler, 2016) include poor teacher content knowledge and pedagogical content knowledge, teachers' beliefs and attitudes, and school structural factors such as lack of time for planning with other teachers. Given the different backgrounds and school contexts that teachers bring to integrated STEM, Vasquez et al. (2013) asserted that STEM integration can start anywhere along the continuum. For example, teachers who have been implementing a disciplinary approach might think about increasing their level of integration to a multidisciplinary approach.

10.3 Methodology

This research project employed design-based research (DBR) methodology (Bannan-Ritland, 2003; Design-Based Research Collective, 2003; van den Akker, Gravemeijer, McKenney, & Nieveen, 2006). The focus of DBR is on refining theoretical understandings through iterative cycles of design, enactment in authentic settings, analysis, and redesign. DBR requires that teachers are prepared and supported in developing a knowledge base sufficient to support the constructive process, both for individual students and for student–student and student–teacher interactions. The design of learning environments is framed by social constructivism theory where students participate individually and collectively through the use of real-world problems.

Research within this study used DBR to progress along a continuum framework of STEM integration (Vasquez et al., 2013) as represented in Fig. 10.1. Researchers and mathematics and science teachers worked together in an attempt to design problems that meaningfully engaged and challenged students to use skills from both disciplines.

10.3.1 Research Context

The host secondary school caters to a diverse student population of around 1,300 students with 46 nationalities represented and international students enrolling each year. The school has a comprehensive academic Victorian Certificate of Education program and a dedicated STEM coordinator working across the middle and senior years. The school has been implementing a STEM project within its curricula and timetable structure since 2017. Three teachers participated in the first cycle of the project in 2017, and two new teachers participated in the second cycle the following year. In 2018, 24 students (11 girls and 13 boys) participated in the study; they worked in eight teams of 2–5 members each. In this chapter, findings from the second cycle will be reported and discussed using teacher interviews from the mid-project and student focus group discussions.

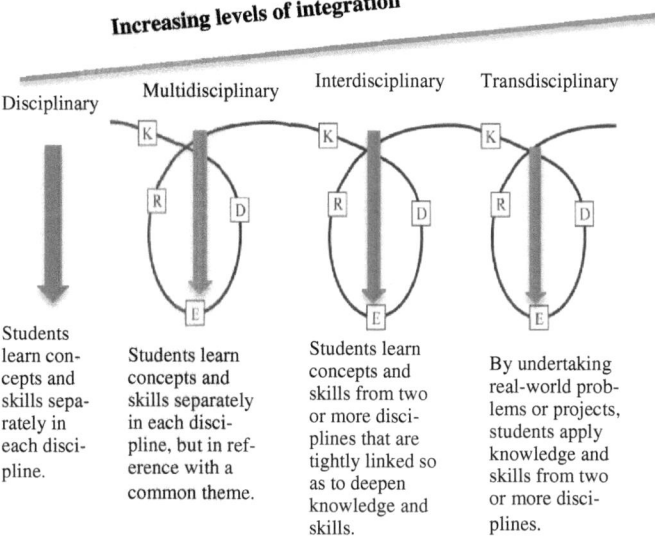

Increasing levels of integration

Disciplinary   Multidisciplinary   Interdisciplinary   Transdisciplinary

| Students learn concepts and skills separately in each discipline. | Students learn concepts and skills separately in each discipline, but in reference with a common theme. | Students learn concepts and skills from two or more disciplines that are tightly linked so as to deepen knowledge and skills. | By undertaking real-world problems or projects, students apply knowledge and skills from two or more disciplines. |

Fig. 10.1 Increasing levels of integration. Adapted from "STEM Lesson Essentials, Grades 3–8: Integrating Science, Technology, Engineering, and Mathematics" by Vasquez et al. (2013), p. 73. Copyright 2013 by Heinemann Publishing Company

10.3.1.1   Description of the Task

The STEM task that was designed by the teachers in collaboration with the researchers was interdisciplinary in nature and authentic in terms of its real-world context. The task related to the context of roller coasters or skateboard parks where students were required to model through the design and creation of a marble run that had certain requirements. The task was framed to integrate STEM disciplines and was designed to address the school curriculum under the science topic Energy in the foreground with the mathematics topic percentages. The preliminary mathematics activity involved students exploring percentages using a suite of research-based interactive computer simulations for teaching and learning physics, chemistry, mathematics, and other sciences (PhET: https://phet.colorado.edu/) on their laptop. In investigating the simulation of a skateboarder in action, the students undertook a percentage calculation of energy changes that took place. Students were asked to represent energy in different ways such as creating a diagram or writing. The preliminary science activity involved students doing experiments with toys (e.g., balls and rubber bands) to explore different types of energy and energy transfer.

The teachers created a common assessment task and rubric to assess student learning outcomes. The key assessment criteria included the following:

- Student's ability t o demonstrate procedural fluency (e.g., being able to calculate a percentage increase given starting and ending points).

- Student's ability to identify, describe, and apply scientific concepts related to energy (e.g., being able to identify and describe the differences between kinetic and potential energy).
- Student's ability to demonstrate skills associated with undertaking an open inquiry (e.g., being able to develop questions about energy to be investigated).
- Student's capacity to demonstrate the twenty-first century skills (e.g., creativity, critical thinking, collaboration, and communication).

The students, working in groups of 2–5, were required to collaborate in the design and creation of the marble run, present their marble run to the rest of the class, and provide a written report as a summative task that addressed a detailed assessment rubric.

10.4   Findings

The school context in which the teachers and students operated is what Clarke and Hollingsworth (2002) referred to as the change environment. According to these authors, the change environment can substantially have an impact on the success of any new initiative that might be undertaken, which in this case was the adoption of an interdisciplinary STEM approach within a real-world context.

While the change environment involved mathematics and science lessons being timetabled separately, when the students were doing the STEM project, the mathematics aspect of the STEM project was addressed during the science lesson and vice versa. To ensure continuity of student progress on the STEM project between the two classes, the mathematics and science teachers updated each other following their lessons. They also developed a detailed task description and assessment rubric that was administered at the completion of the project. The teacher comments that follow are excerpted from their mid-project interviews (MPI).

10.4.1   Potential of Real-World Problems

The application of the real-world context was a significant factor in the success of the project. There was a view that the task teaches mathematics and science concepts that the students applied in a real-world context.

> I think it was the application part of it and the conversations around how to apply the knowledge to – and the skills – to that application … that made it STEM and was … a largely successful part of the project. (Science teacher, MPI)

The real-world context needs to be familiar to the students but not too familiar and still be sufficiently complicated. This view was expressed by the mathematics teacher who thought that the marble run context was appropriate.

It was [a] familiar enough idea that they weren't going to get lost on the context, but it had enough complication and unfamiliarity with what was happening when a marble actually goes down whatever they decided to make it go down. (Mathematics teacher, MPI)

The marble runs not only served as the STEM project but as springboard for other ideas, such as exploring energy changes in a child's toy and investigations of energy changes in a bouncing ball. The science teacher noticed these activities engaged students in exploring scientific ideas related to different types of energy as well as energy transformation.

Some of the [practical work] that we did in terms of tennis balls and bouncing of balls ..., just to get the idea of kinetic gravitational potential ... and then what that energy is turning into when it hits the ground, ... I think that went really well. (Science teacher, MPI)

10.4.2 Awareness of Science, Mathematics, and Technology/Engineering Embedded in the STEM Project

The students were quite comfortable in bringing in the science and mathematics to the given problems. They were aware of the mathematics and science concepts embedded in the task and were able to pinpoint discipline-related concepts they used in the project as evidenced in the student generated reports (SGR) and the student focus group discussion (SFG).

Our marble run sat on an angle and started at 80% and the marble went over a couple of bumps and reached higher than the start, reaching around 100%. Our marble was able to gain enough kinetic energy and friction to roll throughout the whole marble run and land straight in the cup. (SGR)

I put, like, past knowledge that I've learnt in science about gravity and potential energy and kinetic [energy] in the project, to figuring how it's going to work. (SFG)

The teacher noted that students were undertaking more problem-solving, hands-on project than they would have normally undertaken in the classroom.

It [the project] actually gets them up and about and actually use the skills that they've learnt in class to design something. (Science teacher, MPI)

Students enjoyed the hands-on aspects in addition to the discussions with the teacher in exploring ways to solve the problems.

I enjoyed the project quite a lot, it was really interesting, like being able to put your thoughts and actually build something that you want to build. (SFG)

When the marble didn't and did work, we had to find ways to fix or improve the marble run, we had to change our designs to better improve our success rate. (SGR)

The student-centered nature of the project was preferred over the completion of recipe-type experiments and having to follow a series of steps or writing book notes.

[It was] more fun and the teachers get to connect with the students … and you look forward to it more than like learning about something else … instead of like writing in your book. (SFG)

It's more fun and the teachers get to connect with the students. (SFG)

Some students felt that the group composition of friends was distracting at times; this was corroborated by the science teacher.

A couple of students felt that it was good to work with their friends; however, sometimes they can get distracted a little bit more easily and not stay on task but that's, I think, a class matter. (Science teacher, MPI)

However, the STEM project as open-ended and student-centered allowed students to develop their collaborative, creative, and communication skills—all of which are perceived as important to learn and develop.

There was lots of teamwork and skills, like to be able to contribute, everyone's adding to one piece of work actually really worked. (SFG)

I think they're important because, say if one person did all the work then no one would, the rest wouldn't learn anything and it'd be hard to like think of everything by yourself. (SFG)

You could have an open mind and could create different things you wouldn't normally create in science, and we were able to be more independent. (Class presentation)

I think it was good to allow us to use our creativity to build it, and we didn't have to follow any rules. (Class presentation)

10.4.3   Views on STEM Integration

A STEM task needs to involve at least two discipline areas with a practical context to apply ideas from these disciplines. The task needs to be embedded in an inquiry-based approach that is very rich in both small-group and whole-class discussions. This view was articulated by the science teacher.

Well, in terms of STEM as a whole, obviously the concept and the idea around STEM in terms of integrating [2 to 4] different subjects … to come up and to design something that's practical … to apply knowledge that they've learnt in class. (Science teacher, MPI)

Similarly, the integration between mathematics and science in different activities within the larger task was also valued.

They can actually utilize knowledge and try and implement that when they go to do the different activities. So whether that's … playing with toys or building a marble run or utilizing pipes and figuring out what kind of angle they want … to generate or try and keep the marble going. (Mathematics teacher, MPI)

However, from the mathematics teacher's perspective, the STEM project had less mathematics content than science content.

I felt the Maths side has kind of taken a back seat. So, although I've talked about percentages and we put percentages in the task … I've said to the kids, alright give me an example of where one of the following heights is 150% more than the other one, and the kids have kind of then got the idea and it's been good to sort of talk about how percentages are relative, for example, instead of being absolute. (Mathematics teacher, MPI)

The mathematics embedded in the task was explicit where the students needed to create more than one hump in the track with one of the humps being at least 105% higher than the other. What was not made explicit was other mathematics that might be employed, for example, a scale drawing of the model in the design phase.

They're very into drawing … they're just off the top of their head kind of thing and then you've got to convince them to put measurements on and try to get them to think about the energy side of it. (Mathematics teacher, MPI)

Both technology and engineering cycles have specific stages in the construction of the model/product. These are investigating, generating, producing, and evaluating. In this project, the students were seen to rush through the phases of investigating and generating. They were only drawing preliminary sketches to help them in taking materials and constructing their models, which was noticed by the mathematics teacher.

I don't think they really went for any of the thinking-it-through design stuff, they just sort of ticked that box as a token gesture. So, in terms of trying to get them to provide some sort of representation of what they were doing and what they were thinking in terms of when they actually went to go and do the marble run project itself, I don't think they really took that too seriously. (Mathematics teacher, MPI)

The teaching of representations as tools in undertaking the task was important to the topic, and the teachers asserted that students need to pay more serious attention to the design aspect that allows for a deeper discussion of the science concepts.

I understand the representations are a tool, but then if they're not taking the tool seriously or not thinking of it as something that's productive and useful for them, then they're not likely to then use that to go to the next step and then use that properly. (Mathematics teacher, MPI)

There was a view that students need time and practice at the design cycle and that there is a common language and specific representational tools they might employ in addressing a design brief.

They've got a subject now, design technology … and it'd almost be good to look at what other subjects are doing from that point of view with this in mind to see if there is some common language there [so that] we can share with each other and utilize in those sorts of processes even more widely than science and maths as part of the STEM. (Mathematics teacher, MPI)

Implications are that students need appropriate representational tools that are regularly used across the discipline areas.

That's where ideas of representation might come. Each tool needs to be given a label which the students have used and understand its affordances and constraints so when a task is presented to them they can make the decision as to what tool is best. (Mathematics teacher, MPI)

If we were going to take that sort of stuff seriously with the critical thinking and all the rest of it, we probably do need a common set of tools that are regularly used so the students get familiar with them and start using them automatically themselves. (Science teacher, MPI)

10.5 Discussion and Implications

A key driver and facilitator of the interdisciplinary approach was the real-world STEM task. The nature of the task in this case study was well designed in being of a familiar context, which meant that the students could immediately engage with the task, but at the same time was sufficiently open-ended and challenging for them. The STEM task connected the key ideas underpinning the school curricula, namely, percentages (mathematics), energy (science), and the design process (technology). In general, the nature of the real-world task becomes an important aspect of engaging students in interdisciplinary learning of science and mathematics as it enables students to engage in authentic, active, and meaningful learning challenges (Lowrie, Downes, & Leonard, 2018).

Nadelson and Seifert (2017) suggested that, when utilizing integrated STEM contexts, teachers must be aware of the need to identify contextual features that span multiple disciplines and ensure a level of complexity aligned with their students' STEM knowledge and learning capacity. They argued that the success of an integrated STEM approach is determined by the compatibility between the complexity of the task and the students' knowledge level. An integrated context in lessons provides a common experience for students in exploring various aspects of STEM, where students are required to apply knowledge and practices across multiple disciplines. Through undertaking the STEM task, students learned the key concepts of energy transfer, energy transformation, and energy type and applied their skill in working with percentages. The students had completed the mathematical topic of percentages just prior to being introduced to the STEM task. According to Shaughnessy (2013, p. 234), "the M will become silent if not given significant attention" when implementing a STEM task.

This raises the issue as to what represents a good real-world task when implementing an interdisciplinary approach to science and mathematics. Should the science and mathematics be equally represented? Should the curriculum ideas emerge and be learned through undertaking the task or is it sufficient that students apply concepts already learned? We believe that equal weighting and teaching of concepts need not apply in all cases for a claim of interdisciplinary as the connection to and between the disciplines results from the nature of the real-world context. This view concurs with earlier points raised by other researchers (English, 2016, 2017; Shaughnessy, 2013; Treacy & O'Donoghue, 2014) about the need for a STEM task to have genuine integration, that is, where the learning of one discipline does not override others. We acknowledge that in this study achieving equal balance across disciplines particularly for mathematics, engineering, and technology remains a challenge. However, we argue that integrated STEM can be approached from several different perspectives with different disciplines involved in different ways (Hobbs, Cripps Clark, & Plant, 2018) according to the wants and needs of teachers and students and the nature of the real-world task. The exact nature of the benefits and challenges of these different perspectives requires further research.

Apart from addressing the science and mathematics curricula to be taught, the STEM task engaged the students in other science and mathematics topics (i.e., forces [science]; measurement and graphs [mathematics]). While addressing the mathematics and science curricula, the STEM task required the students to engage in a technology/engineering design process and in doing so addressed aspects of the technology curriculum. This raises the issue in that the nature of the real-world task can engage students in several science and mathematics topics and other discipline areas.

Significant benefits lie in students getting insight into (a) the interconnections across multiple concepts within and across discipline areas (English, 2016; English & King, 2015; Shaughnessy, 2013) and (b) the processes of knowledge construction. Within this study, the STEM task required that students practice and develop their skills to problem solve, scientifically investigate, and engage in technology/engineering design (English & King, 2015; Marginson et al., 2013; Wang, Moore, Roehrig, & Park, 2011). Such process skills are very much part of the curriculum as are concepts and provide students with insights into the daily lives of scientists, mathematicians, engineers, and technologists (Leung, 2018; Lowrie et al., 2018).

Within this study, the real-world task and the change environment facilitated and constrained the nature of the interdisciplinary approach undertaken by the mathematics and science teachers. Clarke and Hollingsworth (2002) suggested that any change in a teachers' professional practice occurs within the constraints and affordances of the change environment. They argued that school context

> can impinge on a teacher's professional growth at every stage of the professional development process: access to opportunities for professional development; restriction or support for particular types of participation; encouragement or discouragement to experiment with new teaching techniques; and, administrative restrictions or support in the long-term application of new ideas. (Clarke & Hollingsworth, 2002, p. 962)

Mathematics and science were timetabled separately, which meant that the teachers needed to collaborate closely in both the planning and delivery stages. This is consistent with the point of Vasquez (2014, p. 15) that "Developing integrated STEM experience is not a linear process. It takes collaboration and preparation." Continuity for the students was realized through having to undertake the same task in both their science and mathematics lessons and for them to address the one assessment rubric, which not only addressed the science and mathematics concepts and skills but also their technology/engineering design skills and general STEM competencies such as creativity and teamwork. The real-world problem transcended the individual disciplines of science and mathematics; therefore, students were undertaking a STEM project whether it occurred in a timetabled mathematics class o r sience class. The STEM project drove the engagement and learning for the students—not the disciplines of mathematics or science.

In general, an interdisciplinary approach to teaching mathematics and science should not be seen as one particular practice. Instead, it should be seen as a part of a continuum of practices that connect STEM disciplines within a real-world problem and connect the mandated curriculum within the constraints of the change environment. Successful STEM integration requires teachers to see and experience an integrated STEM task as enhancing and not adding to the existing curriculum (Bryan et al., 2015).

Research has indicated that factors involved in professional learning that lead to effective integrated STEM programs include collective participation, active learning, content knowledge, coherence, and duration (Johnson & Sondergeld, 2015). In terms of coherence and duration, it is vital that planning the STEM initiative from the perspective of the teachers be consistent with the school's priorities. There was clear importance placed on significant contact hours (at least 80 h) in developing strategies to get insights into the tradition of mathematics for science teachers and vice versa for the mathematics teachers. While these strategies focused on working within the professional learning context, it highlights the importance of planning between teachers.

Within this study, the science and mathematics teachers collaborated in the planning, delivery, and assessment of the STEM task. This took time; given that schools often only provide teachers time to plan and collaborate within their discipline area, an interdisciplinary approach requires a change in school culture as to the provision in the timetable for interdisciplinary planning. In addition, these mathematics and science teachers saw the need to learn more about the design process as espoused by technology teachers. Schools should provide time for teachers to learn key elements across each STEM discipline. This may lead to more efficient ways to address the overall curriculum and common pedagogical approaches with a similar language used by all the science teachers. One challenge of teaching integrated STEM in Australia is that the Australian educational system is discipline based; moving toward problem-based learning would require restructuring of planning and implementation of the curriculum. Nadelson and Seifert (2017) pointed out that reconciling "the historical structure of the schools, curriculum instruction, and assessment to create a school culture and environment that supports an integrated STEM approach to teaching and learning" (p. 223) needs to be addressed. A successful interdisciplinary approach requires a school-wide perspective and support from teachers, administrators, and students (Nadelson & Seifert, 2017; Sanders, 2012) and focuses on learning outcomes for students (Siekmann & Korbel, 2016; Vasquez, 2014).

Acknowledgements We would like to thank the teachers and students in the school who contributed to this study and Dr. Esther Y.-K. Loong for her contribution to the project. The study was funded by Research for Educational Impact, Faculty of Arts and Education, Deakin University, Australia.

References

Bannan-Ritland, B. (2003). The role of design in research: The integrative learning design framework. *Educational Reseacher, 32*(1), 21–24.

Blum, W., & Niss, M. (1991). Applied mathematical problem solving, modelling, applications and links to other subjects. *Educational Studies in Mathematics, 22*(1), 37–68.

Bryan, L. A., Moore, T. J., Johnson, C. C., & Roehrig, G. H. (2015). Integrated STEM education. In C. C. Johnson, E. E. Peters-Burton, & T. J. Moore (Eds.), *STEM road map: A framework for integrated STEM education* (pp. 23–37). New York, NY: Routledge.

California Department of Education. (2014). *Innovate: A blueprint for science, technology, engineering, and mathematics in California public education* (STEM Task Force Report). Sacramento, CA: Author. Available from https://www.cde.ca.gov/pd/ca/sc/documents/innovate.pdf.

Clarke, D., & Hollingsworth, H. (2002). Elaborating a model of teacher professional growth. *Teaching and Teacher Education, 18,* 947–967.

Corlu, M. S., Capraro, R. M., & Capraro, M. M. (2014). Introducing STEM education: Implications for educating our teachers for the age of innovation. *Education and Science, 39*(171), 74–85.

Design-Based Research Collective. (2003). Design-based research: An emerging paradigm for educational inquiry. *Educational Researcher, 32*(1), 5–8.

Dreher, A., Kuntze, S., & Lerman, S. (2016). Why use multiple representations in the mathematics classroom? Views of English and German preservice teachers. *International Journal of Science and Mathematics Education, 14*(2), 363–381. https://doi.org/10.1007/s10763-015-9633-6.

English, L. D. (2009). Promoting interdisciplinarity through mathematical modelling. *ZDM Mathematics Education, 41,* 161–181.

English, L. D. (2016). STEM education K-12: Perspectives on integration. *International Journal of STEM Education, 3*(1). https://doi.org/10.1186/s40594-016-0036-1.

English, L. D. (2017). Advancing elementary and middle school STEM education. *International Journal of Science and Mathematics Education, 15*(Suppl. 1), S5–S24.

English, L. D., & King, D. T. (2015). STEM learning through engineering design: Fourth-grade students' investigations in aerospace. *International Journal of STEM Education, 2*(1). https://doi.org/10.1186/s40594-015-0027-7.

Ferri, R. B., & Mousoulides, N. (2017). *Mathematical modelling as a prototype for interdisciplinary mathematics education? Theoretical reflections.* Paper presented at the CERME 10, Dublin, Ireland.

Funner, J. M., & Kumar, D. D. (2007). The mathematics and science integration argument: A strand for teacher education. *Eurasia Journal of Mathematics, Science & Technology Education, 3*(3), 185–189.

Gardner, M., & Tillotson, J. W. (2018). Interpreting integrated STEM: Sustaining pedagogical innovation within a public middle school context. *International Journal of Science and Mathematics Education.* Advance online publication. https://doi.org/10.1007/s10763-018-9927-6.

Hobbs, L. (2012). Teaching out-of-field: Factors shaping identities of secondary science and mathematics. *Teach Science, 58*(1), 21–29.

Hobbs, L., Cripps Clark, J., & Plant, B. (2018). Successful students—STEM program: Teacher learning through a multifaceted vision for STEM education. In R. Jorgensen & K. Larkin (Eds.), *STEM education in junior secondary* (pp. 133–168). Singapore: Springer.

Honey, M., Pearson, G., & Schweingruber, H. (Eds.). (2014). *STEM integration in K-12 education: Status, prospects and an agenda for research.* Washington, DC: National Academies Press.

Johnson, C. C., & Sondergeld, T. A. (2015). Effective STEM professional development. In C. C. Johnson, E. E. Peters-Burton, & T. J. Moore (Eds.), *STEM road map: A framework for integrated STEM education* (pp. 203–210). New York, NY: Routledge.

Johri, A., Roth, W.-M., & Olds, B. (2013). The role of representations in engineering practices: Taking a turn towards inscriptions. *Journal of Engineering Education, 102*(1), 2–19. https://doi.org/10.1002/jee.20005.

Latour, B. (1999). *Pandora's hope. Essays on the reality of science studies*. Cambridge, MA: Harvard University Press.

Lehrer, R., & Schauble, L. (2000). Developing model-based reasoning in mathematics and science. *Journal of Applied Developmental Psychology, 21*(1), 39–48.

Lehrer, R., Schauble, L., & Lucas, D. (2008). Supporting development of the epistemology of inquiry. *Cognitive Development, 23*(4), 512–529.

Leung, A. (2018). Exploring STEM pedagogy in the mathematics classroom: A tool-based experiment lesson on estimation. *International Journal of Science and Mathematics Education*. Advance online publication. https://doi.org/10.1007/s10763-018-9924-9.

Lowrie, T., Downes, N., & Leonard, S. (2018). *STEM education for all young Australians: A Bright Spots STEM Learning Hub Foundation Paper for SVA, in partnership with Samsung*. Canberra, ACT: University of Canberra STEM Education Research Centre. Retrieved from http://www.socialventures.com.au/assets/STEM-education-for-all-young-Australians-Smaller.pdf.

Marginson, S., Tytler, R., Freeman, B., & Roberts, K. (2013). *STEM: Country comparisons—International comparisons of science, technology, engineering and mathematics (STEM) education*. Melbourne, Australia: Australian Council of Learned Academies. Retrieved from https://acola.org.au/wp/PDF/SAF02Consultants/SAF02_STEM_%20FINAL.pdf.

Moore, T. J., Glancy, A. W., Tank, K. M., Kersten, J. A., Smith, K. A., & Stohlmann, M. S. (2014). A framework for quality K-12 engineering education: Research and development. *College of Pre-College Engineering Education, 4*(1). https://doi.org/10.7771/2157-9288.1069.

Nadelson, L. S., & Seifert, A. L. (2017). Integrated STEM defined: Contexts, challenges, and the future. *Journal of Educational Research, 110*(3), 221–223.

Pearson, G. (2017). National academics piece on integrated STEM. *Journal of Educational Research, 110*(3), 224–226.

Ríordáin, M. N., Johnston, J., & Walshe, G. (2016). Making mathematics and science integration happen: Key aspects of practice. *International Journal of Mathematical Education in Science and Technology, 47*(2), 233–255.

Sanders, M. E. (2012). Integrative STEM education as best practice. In H. Middleton (Ed.), *Explorations of best practice in technology, design, & engineering education* (Vol. 2, pp. 103–117). Queensland, Australia: Griffith Institute for Educational Research.

Schoenfeld, A. (2004). Multiple learning communities: Students, teachers, instructional designers, and researchers. *Journal of Curriculum Studies, 36*(2), 237–255.

Shaughnessy, J. M. (2013). Mathematics in a STEM context. *Mathematics Teaching in the Middle School, 18*(6), 324–327. https://doi.org/10.5951/mathteacmiddscho.18.6.0324.

Siekmann, G., & Korbel, P. (2016). *Defining 'STEM' skills: Review and synthesis of the literature*. Adelaide, Australia: National Centre for Vocational Education Research (NCVER). Available from https://www.ncver.edu.au/research-and-statistics/publications/all-publications/what-is-stem-the-need-for-unpacking-its-definitions-and-applications.

Stinson, K., Harkness, S. S., Meyer, H., & Stallworth, J. (2009). Mathematics and science integration: Models and characterizations. *School Science and Mathematics, 109*(3), 153–161.

Treacy, P., & O'Donoghue, J. (2014). Authentic integration: A model for integrating mathematics and science in the classroom. *International Journal of Mathematics Education in Science and Technology, 45*(5), 703–718.

Tytler, R. (2016, July 24–31). *Challenges for mathematics within an interdisciplinary STEM education*. Paper (keynote) presented at the 13th International Congress on Mathematical Education. Hamburg, Germany.

van den Akker, J., Gravemeijer, K., McKenney, S., & Nieveen, N. (2006). *Educational design research*. Abingdon, England: Routledge.

Vasquez, J. A. (2014). STEM beyond the acronym. *Educational Leadership,* (December), 10–15.

Vasquez, J. A., Sneider, C., & Comer, M. (2013). *STEM lesson essentials, Grades 3–8: Integrating science, technology, engineering, and mathematics*. Portsmouth, NH: Heinemann.

Wang, H.-H., Moore, T. J., Roehrig, G. H., & Park, M. S. (2011). STEM integration: Teacher perceptions and practice. *Journal of Pre-College Engineering Education Research, 1*(2). https:// doi.org/10.5703/1288284314636.

Williams, J., Roth, W.-M., Swanson, D., Doig, B., Groves, S., Omuvwie, M., et al. (2016). *Interdisciplinary mathematics education: A state of the art*. Cham, Switzerland: Springer.

Chapter 11
Framing and Assessing Scientific Inquiry Practices

Russell Tytler and Peta White

11.1 Introduction

Inquiry as a core element of science curricula has a long history in the ideas of educators like Schwab (1962, 1964) and Dewey (1996). Schwab (1962) argued for the importance of representing the way science ideas are developed and validated as against the prevailing strong focus on the products of scientific knowledge, which he famously characterized as a "rhetoric of conclusions". Osborne (2006) pointed out that:

> Four decades after Schwab's (1962) argument that science should be taught as an 'enquiry into enquiry', and almost a century since John Dewey (1916) advocated that classroom learning be a student-centred process of enquiry, we still find ourselves struggling to achieve such practices in the science classroom. (p. 2)

The literature is replete with descriptions of traditional science teaching and learning as consisting largely of teacher presentation and closed questioning that pays insufficient attention to the development of higher thinking processes (Edwards & Mercer, 1987; Goodrum, Hackling, & Rennie, 2001). Similarly, practical work largely consists of explicit instructions that reduce the experience of students to procedural inquiry protocols, observations, and results intended to be illustrative of knowledge already determined (Hassard & Dias, 2008; Holmes & Wieman, 2016). In this case, students are not given access to the investigative challenges that would expose them to higher order thinking and decision-making and leave untouched the misleading empiricist epistemology that posits experimentation as determining scientific insights in a non-problematic relationship (Carey, Evans, Honda, Jay, & Unger, 1989).

One of the problems in talking or writing about inquiry is that it covers a multitude of approaches to teaching and learning in science (Anderson, 2002; Chen & Tytler,

R. Tytler (✉) · P. White
Faculty of Arts and Education, Deakin University, Geelong, Australia
e-mail: tytler@deakin.edu.au

© Springer Nature Singapore Pte Ltd. 2019
Y.-S. Hsu and Y.-F. Yeh (eds.), *Asia-Pacific STEM Teaching Practices*,
https://doi.org/10.1007/978-981-15-0768-7_11

2017). It is understood broadly as a pedagogy that has students posing questions and exploring ideas prior to teacher explanation. The focus in this pedagogy is on higher level thinking and reasoning. A more specific but important aspect of inquiry involves students in practical investigations that focus on the specific processes by which scientific knowledge is built through empirical evidence.

Recently, there has been increasing recognition of the need to develop students' knowledge of the processes by which scientific knowledge is generated and validated—epistemic knowledge or knowledge of the epistemic processes of science (Duschl, 2008). An important aspect of scientific literacy is knowledge of evidential processes—the process by which theories are generated and tested. A substantial research interest in argumentation has grown out of this concern (Simon, Erduran, & Osborne, 2006). In our own research drawing on increasing recognition of the importance of multimodal representations in the generation of scientific knowledge (Gooding, 2004; Latour, 1999) and the importance of modeling processes as the basis for scientific knowledge building (Lehrer & Schauble, 2012), we emphasize in our inquiry work the importance of students' imaginative construction of explanatory representations and models as part of their classroom activities (Tytler, Prain, Hubber, & Waldrip, 2013).

11.2   Curriculum Framing of Inquiry Processes

Recent calls for science education reform have focused on the need to have students develop the sets of skills that will prepare them for a complex future in their lives and work. Such twenty-first century skills include complex problem-solving, critical and creative reasoning, collaboration, communication, and digital literacy. Contemporary science curricula increasingly emphasize inquiry and higher order skills and understandings o f he nature of scientific endeavor, for instance in the USA through the *Next Generation Science Standards* (National Research Council, 2013), in Australia through an inquiry skills strand (Australian Assessment, Curriculum, and Reporting Authority [ACARA], 2013), and in Taiwan through the *Curriculum Guideline of Science and Technology* (Ministry of Education, 2013).

Further, current emphases in science, technology, engineering, and mathematics (STEM) education foreground the importance of teaching and learning through authentic problem-based contexts that bring science into contact with mathematics, engineering, and digital technologies. These demands place an extra burden on the identification of core scientific processes that can be flexibly applied—and assessment approaches that clarify—and support these.

Yet, alongside these emphases on higher level inquiry and problem-solving skills applied in context, there is an increasing emphasis on accountability measures for curriculum and teaching and learning that include standardized and decontextualized testing and outcome tracking. These trends are in many respects contradictory, in that the demands for reliable and valid testing regimes tend to preclude assessments that are complex and/or contextual as higher order competencies tend to require.

There is a history of development of performance assessments (Palm, 2008) that we argue are effective but limited in scope for encouraging or measuring students' capacity for responding to the complexity inherent in real contexts. System-wide performance assessments tend to be very expensive in terms of development, administration time, and results analysis. Moreover, trends in educational assessment emphasize a more complex and generative role for assessment than the traditional summative role, with assessment *for* and *as* learning gaining in importance as a foundation for responsive teaching and learning that can help teachers articulate and develop students' inquiry skills.

> It has become common in education to refer to the multiple 'purposes' of assessment. But a conceptual breakthrough is made by recognising that there is only one fundamental purpose of assessment in education. That purpose is to establish and understand where learners are in an aspect of their learning at the time of assessment. (Masters, 2015, p. 2)

Thus, given this focus on assessment for formative purposes, there is a need to develop inquiry skills assessment that is contextual and school based and to assess and support students' inquiry skill development.

Existing traditions in science teaching and learning, including in practical work, tend to emphasize teacher-dominated knowledge delivery pedagogies and set piece, procedurally focused, illustrative, practical work. Such a tradition has epistemological and pedagogical entailments that dominate science teaching and learning cultures. Developing effective assessment regimes is, therefore, important for the professional learning of teachers so as to establish a language and pedagogy that support the development of the higher order inquiry processes. Such situated, school-based assessments of inquiry skills, should

1. Support assessment of inquiry skills *for* and *as* learning, carried out as part of core inquiry teaching and learning processes;
2. Represent contemporary perspectives on the nature of scientific inquiry;
3. Emphasize higher level performance, including problem-solving, model-based reasoning, and the coordination of ideas with evidence; and
4. Support teacher epistemological and pedagogical development that will enable a sharper and more pervasive focus in classroom practice on inquiry skills and epistemic knowledge.

In this chapter, we describe two Australian initiatives aimed at developing instruments that incorporate targeted practical activity to support and assess students' inquiry skill development. In neither case has a fully validated instrument been developed, but our aim is to describe the nature of the approach, teacher adaptations of the instruments, and teacher perceptions and practice in order to lay out the principles and challenges involved in assessing inquiry in context.

11.3   The Cases

The first initiative is a lower secondary school project—the Victorian node of
the Australian Advancing Science and Engineering through Laboratory Learning (ASELL) for Schools project (https://blogs.deakin.edu.au/asell-for-schools-vic/
asell-for-schools/)—that involved the development of practical laboratory activities
for schools. Experience from this project led to the Victorian team developing an
approach to inquiry informing practical work design with an explicit focus on both
contemporary science and student representational work.

The second initiative was funded by the Victorian Department of Education and
Training as part of a major Primary Mathematics and Science Specialist (PMSS) program. It involved the refinement and informal validation of science inquiry assessment activities, which was used to evaluate the wider program aimed at educating
primary science specialists.

In both cases, the project materials and assessment were aligned with the inquiry
skills strand of the Victorian Curriculum (Victorian Curriculum and Assessment
Authority, 2015), which bears a close resemblance to the Australian Curriculum
(ACARA, 2013) outcome framework. The inquiry skills framework in the Victorian
Curriculum can be seen in the first column of the Inquiry Scaffold Tool (Fig. 11.1;
Deakin University, n.d.). In each case, one or more of the authors was involved in
developing the instruments and investigating teacher responses to them.

11.3.1   Case 1: The Victorian Node of ASELL
for Schools—Inquiry Scaffold Tool

The ASELL for schools project was national (i.e., taken up in many Australian
states), funded under the Australian Mathematics and Science Partnership Program,
and aimed to develop or redevelop laboratory learning activities (LLA). At the heart
of an LLA is the understanding that *science is everywhere* and that science activities
should guide students into deeper conceptual understanding while also developing
inquiry skills to replicate how scientists conduct research. The Victorian Node of the
ASELL for Schools (ASELL-SVN) project added additional dimensions to engage
with local educational priorities and practices, including the three curriculum strands:
science understandings, science as a human endeavor, and science inquiry skills. The
project worked to design new, or redesign existing, LLAs that incorporate these three
strands in innovative ways.

Five principles framed the development of each LLA. First, the use of *inquiry-
based learning* to model scientific practices for students and give them increasingly
more independence in all steps of the scientific process. Second, attempts to engage
students with *conceptual learning* that successfully link hands-on laboratory activities to conceptual knowledge, helping the student make deeper connections. Third,

SCHOOLS
Advancing Science & Engineering through Laboratory Learning

Curriculum outcome - Years 7-8	Prescription	Confirmation	Structured Inquiry	Guided Inquiry	Open Inquiry
Questioning and predicting Identify questions, problems and claims that can be investigated scientifically and make predictions based on scientific knowledge (https://victoriancurriculum.vcaa.vic.edu.au/Curriculum/ContentDescription/VCSIS107)	Student engages with a question provided by teacher.	Student chooses from a provided, constrained set of questions.	Student sharpens or clarifies a question or questions provided by teacher, or other source.	Based on discussion with teacher, or others, student poses and refines their own question.	Student autonomously poses a question of interest.
Planning and conducting Collaboratively and individually plan and conduct a range of investigation types, including fieldwork and experiments, ensuring safety and ethical guidelines are followed (https://victoriancurriculum.vcaa.vic.edu.au/Curriculum/ContentDescription/VCSIS108)	Student follows a provided plan of investigation.	Student follows a plan that offers limited choices in approach.	Student adapts and refines a provided plan outline.	Student uses a planning framework to devise and enact a plan.	Student autonomously devises and enacts a plan for a chosen investigation.
Recording and processing Construct and use a range of representations including graphs, keys and models to record and summarise data from students' own investigations and secondary sources, and to represent and analyse patterns and relationships (https://victoriancurriculum.vcaa.vic.edu.au/Curriculum/ContentDescription/VCSIS110)	Student uses provided representations such as tables to record / process data.	Student chooses from provided representations to record / process data.	Student draws on a structured framework to construct representations to record / process data.	Student draws on a provided outline to develop representations to record / process data.	Student autonomously develops representations to appropriately record / process data.
Analysing and evaluating Use scientific knowledge and findings from investigations to identify relationships, evaluate claims and draw conclusions (https://victoriancurriculum.vcaa.vic.edu.au/Curriculum/ContentDescription/VCSIS111)	Student is strongly directed towards a conclusion.	Student follows instructions involving limited decisions on analysing / concluding.	Student draws on a framework suggesting approaches to analysing / concluding.	Student develops approaches to analysing / concluding, supported by a provided outline.	Student autonomously develops approaches to analysing data and drawing conclusions.

Fig. 11.1 The Inquiry Scaffold Tool for years 7–8 of the Victorian Curriculum as developed by the ASELL-SVN academics in collaboration with many engaged Victorian teachers (Deakin University, n.d.)

the use of the *representation construction approach* to give students the opportunity to demonstrate and develop their knowledge not only with words but also with multimodal representations (e.g., diagrams, roleplays, animations). Fourth, a focus on the *assessment of inquiry skills* to provide teachers with encouragement to focus on one or two of the inquiry skill outcomes in each LLA. This builds students' skills toward being able to conduct open-ended investigations, which is required at the upper secondary school level in Victoria. The Inquiry Scaffold Tool (explained below) is introduced and used to scaffold teaching and learning for each inquiry skill outcome. Finally, a focus on incorporating *contemporary science practice* to make science contextual and relevant to students and help them link conceptual knowledge learned in the classroom to real-life applications of science, often through current research projects representing scientists and their data.

To illustrate the application of the principles, an example of one LLA is offered. This activity was one of five developed in conjunction with the project called Introducing Modern Materials. This project arose from a collaboration between the Reconceptualising Mathematics and Science Teacher Education Project (http://remstep. org.au/), the Deakin University Institute of Frontier Materials (IFM), and ASELL-SVN. Modern materials have changed drastically over the last 50 years with the development of communications technologies such as mobile phones or tablets, new fabrics including sportswear with carefully engineered properties, lightweight vehicles, and space-age technologies such as nanotechnologies and smart materials. The activities introduced a number of important ideas related to modern materials: (a) the nature of modern materials and their uses; (b) the practices of scientists, engineers, and mathematicians at IFM as they invent and produce these materials; (c) what it is like to be a scientist or engineer at IFM; and (d) how to think about the properties of modern materials and how they work.

The website offers this activity and additional material can be seen at this address https://blogs.deakin.edu.au/remstep/. The specific activity described here is called composite materials and can be found at this website https://blogs.deakin.edu.au/ remstep/materials-activities/composite-materials/. The activity enables student to experience and investigate the way combining the properties of two materials (Styrofoam and duct tape) can result in a composite material of different properties, in this case, significantly increased strength. During the process, the students learn about tension and compression forces. This LLA unpacks how composite materials can respond differently to unbalanced forces. The LLA encourages students to design and conduct a strength test, which is a common engineering investigation. Much of the language used includes engineering terminology and in real ways engages students in the everyday practices of an engineer. The use of engineering investigations and language is an important part of STEM education that is often not considered. Curriculum outcomes explored through this activity include science as human endeavor, science understandings—physical science, design and technology outcomes in technological contexts, and science inquiry skills.

The ASELL-SVN project offered workshops hosted by teachers, who invited colleagues and students to join for the day. During each workshop, two LLAs were trialed and feedback sought with the aim of improving the documentation and activity. Both students and teachers provided valuable insights. The students then participated in a hands-on session with a research scientist or engineer while the teachers participated in a pedagogically focused session. It was during the conduct of the teacher sessions that the ASELL-SVN academics realized the need to focus explicitly on the assessment of inquiry skills.

Disappointingly, in a survey canvassing teacher responses regarding their confidence with inquiry teaching and assessment, very few teachers (out of the ~400 surveyed) indicated that they felt competent and satisfied with their teaching practice. This prompted the development by the ASELL-SVN team of the Inquiry Scaffold Tool (Fig. 11.1; Deakin University, n.d.).

The scaffold tool is an extension of previous schemes (National Institutes of Health, 2005; National Research Council, 2000) to conceptualize different degrees of openness in inquiry investigations, from prescribed and teacher directed activity through to open and student-led activity, across different dimensions of investigation. The tool was refined through an iterative process of development by the ASELL-SVN team and trialing in discussion with teachers at workshops over 3 years. The refinement process involved team meetings where teacher feedback was analyzed and the scaffold tool descriptors refined to offer a coherent progression for each skill. The movement across levels is based on increasing independence and sophistication of student inquiry processes, which is consistent with the growth in skills described by the Victorian Curriculum.

The website introducing the Inquiry Scaffold Tool (Deakin University, n.d.) makes the following points:

- The tool offers a conception of the way that teachers provide students with inquiry scaffolding at different levels. Across the levels in the tool, the dimension of change is the degree of agency and responsibility accorded the student for making informed decisions and exhibiting independent inquiry skills. At the prescription level, the teacher strongly frames inquiry, and models the skills through direction;
- The tool offers a way of thinking about the degree of scaffolding put around each skill, with support being reduced at each successive level;
- Teachers should focus on one or two of the seven inquiry skills in each laboratory learning activity; and
- Teaching inquiry skills necessitates direct teaching and skill learning prior to assessing the development of each skill. (para. 2)

Practical activities are designed so that for any particular inquiry skill, the level of scaffolding can be described according to a developmental progression:

Prescription: The student performs the skill strongly scaffolded by explicit instructions. This might involve a highly directive worksheet, or teacher instruction.

Confirmation: The student makes constrained choices within a set of instructions, or strongly guided class discussion. There is minimal room for variation.

Structured inquiry: The student interprets and modifies inquiry processes within an explicit framework. This may involve prior class discussion.

Guided inquiry: The student is involved in substantial decision making and interpretation within a broad outline of suggestions of possible approaches.

Open inquiry: The student engages with a question or problem that they have posed and are invested in, and conducts an investigation with minimal guidance. (para. 4)

There is flexibility built into the Inquiry Scaffold Tool. For instance,

even if an inquiry was intended to develop the open inquiry level of a skill, in supporting individual students the teacher would provide [targeted] guidance characteristic of the lower levels. The tool therefore supports the application of individualized learning and [differentiated support and assessment.] (para. 5).

In the ASELL-SVN project, teachers are encouraged to not start again with designing learning activities for students, but to redesign existing practical activities, adapting the focus more explicitly toward the teaching and learning of the inquiry skills. In this task the Inquiry Scaffold Tool is used to map the outcome focus with each activity through the identification of which inquiry skills are most appropriately addressed through the LLA. The task is then to identify where to decrease the inquiry skill scaffolding, making it appropriate for both the context and the lesson's intended learning outcomes.

As part of a school's program of students' inquiry skill development, it is envisaged that the Inquiry Scaffold Tool (Deakin University, n.d.) could be used in a number of ways:

- To plan a structured program to support the development of individual inquiry skills;
- To map the inquiry skill outcome for each practical activity and provide suggestions for differentiation of student learning;
- To map all inquiry skill outcomes across a unit or year level, scaffolding the development of each skill; and
- To map inquiry skills across all years, building student capacity toward the open investigations found in the senior secondary sciences. (para. 6)

The Inquiry Scaffold Tool is thus designed with two major purposes in mind: to support teachers to design inquiry tasks (e.g., LLAs) that focus explicitly on skill development, and to support the staged assessment of inquiry skills. For the latter purpose, as an example, a teacher might design or use an LLA that is at a guided inquiry level for the analyzing and evaluating skill then providing through instructions or class discussion the principles of analysis. Students, in carrying out the LLA, may then operate at lower levels, needing different amounts of teacher suggestions or support to complete the task successfully. Assessment then involves recording the level of support needed, together with judgment of the level of performance evidenced in students' reports.

11.3.2   Case 2: The Science Inquiry Assessment Tool

The science inquiry assessment (SIA) was conceived as part of the evaluation process for the PMSS program, which involved 40+ teachers training to be specialists and supporting science reform in their schools. The pedagogies emphasized in the program were broadly inquiry-based. Given the lack of a state-wide science assessment regime and the degree of topic choice in the curriculum, it made sense that the SIA represented the major aspect of evaluation of student outcomes.

A first draft of the inquiry assessment tasks, developed in-house by the Department of Education and Training, was trialed with schools involved in the Science Specialist program in 2012. Schools in that first year provided feedback on task appropriateness, language, and task rubrics. Schools reported spending excessive time carrying out the assessment for some tasks and called for tasks to be more streamlined and adjusted for consistency to grade levels. A review and redevelopment were carried out by a Deakin University team, managed by the first author, in preparation for a 2014 trial. Items that required more than one hour of class time were redesigned with attention to equipment demands, the rubrics reviewed and refined, and replacement tasks developed to ensure that each pair of assessments (pre and post) were equivalent. The instrument consisted of two tasks for each year level, generally with three inquiry skills assessed. The two tasks were adjusted to be equivalent in skill focus and type of activity. One task was to be administered in the early part of the year (March) and the other in the later part of the year (September) to measure growth.

The subsequent refinement and informal validation of the instrument involved a number of processes. First, in a session with the science specialists, groups worked with one assessment task each, exploring the appropriateness of the activity for the intended year level. Specialists were asked to discuss what sort of responses would be expected so as to match the rubric descriptions. Groups then provided feedback on the age-appropriateness of the task, the practicalities of running the task for the teacher, and the clarity of the rubric.

Figure 11.2 shows a sample task for Grade 4 (9-year-old children) that focused on the inquiry skills of questioning and predicting, planning and conducting, and evaluating. The instrument consists of (a) a task description that prescribes the actions of teachers in setting up the task, supporting students, and recording responses and (b) a rubric that gives descriptors at progression points spaced at half-year intervals. The rubric consists of general descriptors, refined by the team, and also a descriptor interpreting it for the specific activity. The instrument was refined and used by schools during 2014. The challenges we faced in developing and refining the instrument are illustrative of inquiry assessment processes. These were:

- Designing tasks that were practical in the time available, were age appropriate, and allowed a continuum of responses across the target skills.
- Providing clarity of instructions to teachers who may not be fully familiar with the science concepts or may not possess experience in recognizing the different levels of inquiry skill. We were conscious that the instrument potentially represented a significant professional learning opportunity for schools and teachers.
- Making decisions about how to prescribe different levels of intervention for teachers that would balance a need to leave students room to display skills independently but provide scaffolding for students who could display the skill with different levels of support.

Task Level 4b: **Dissolving Sugar Investigation** *(September)*

<u>Question for investigation</u>: Which type of sugar (normal sugar, raw sugar, coffee crystals and a sugar cube) will dissolve the quickest?

<u>Assessment focus</u>: This task focuses on children's ability to generate testable questions and sensibly predict outcomes, plan and conduct an investigation involving variable control and appropriate measurement processes, and reflect on how they could improve their methods.

<u>Preparation: Each group should have a</u> stirrer, stopwatch, containers of normal sugar, raw sugar, coffee crystals and a sugar cube, containers (e.g. plastic cups) for mixing, spoon or scales.

<u>Task</u>: The investigation requires students to design then carry out an investigation in small groups identifying and controlling variables.

Demonstrate dissolving sugar in a glass jar, discussing with students the meaning of 'dissolving' and raising questions – how can we know when the sugar has finally dissolved? Discuss children's ideas about what might affect the rate of dissolving and why. Discuss prior expectations of which type of sugar will dissolve quickest and why– crystal size, hardness, surface area etc.

Groups then plan an investigation to test which sugar dissolves fastest. What other things might affect how quickly sugar dissolves? Groups construct a list of what they need to keep the same. They discuss how they are going to record results.

Discuss these ideas with the class, and discuss fair tests, and how all other variables need to be controlled (amount of water, amount of sugar, temperature, stirring method). The types of sugar are different. What will they need to keep the same?

Students will plan and carry out the investigation and record results.

<u>Task</u>: Students compare their result with other groups? Did everyone get the same result? Ask each group: was your test fair? How could you improve it?

Students write a report of what they found, and their reflection.

Fig. 11.2 a Sample inquiry skills assessment task (For Grade 4 level). **b** Assessment rubric for the inquiry skills task shown in Fig. 11.2a

Assessment Rubric

Instructions: This rubric is for the teacher to use when working through the task. Highlight the skill level at which the student is achieving in the three areas.

	Curriculum Progression Point			
	3.0	3.5	4.0	4.5
Questioning and predicting	With guidance, poses questions in familiar and structured contexts and suggests ways to investigate these, and predictions of outcomes. *With guidance suggests questions about why the different sugars may dissolve differently.*	Generates questions that can be tested with simple teacher-led investigations *Generates questions and speculations about why the different sugars may dissolve differently*	With guidance, identifies questions in familiar contexts that can be investigated scientifically, and makes predictions based on prior knowledge *Generates questions about different conditions and predicts outcomes.*	With guidance from the teacher or others, poses questions appropriate for scientific investigation and makes predictions and justifies these. *With support articulates questions in a way that can be tested and relates these to scientific ideas.*
Planning and conducting	Within teacher guided discussions, suggests and plans ways of investigating the answers to questions, and uses appropriate tools to make and record observations, including some use of formal measurements *With prompting can sensibly discuss plans and follow guidance in making and recording observations.*	*With the support of the group plans and conduct the investigation, and record results.*	Suggests ways to plan and conduct investigations to answer questions. Identifies the variables in an experiment and with support plans for variable control *Suggests ways to compare sugars and recognises variables on prompting. Independently makes and records times.*	*Can plan and conduct a fair test with minimal guidance.*
Evaluating	Suggests reasons for differences in findings and considers how to improve investigative methods.	*With guidance identifies where improvements to their investigation could be made.*	Reflect on their investigation including suggests reasons why methods were fair or not	*Independently describes ways to improve the fairness of their investigation.*

Fig. 11.2 (continued)

- Creating rubric wording that was distinct for each progression point, avoiding as much as possible resorting to adjectives that describe different degrees of the same performance, and interpreting these generic skill levels in the particular case of the task. We were conscious that this involved a considerable degree of on-balance judgment.

The instrument was used by the 34 schools involved in the project (involving 5,861 matched student results) to track progress across a 6-month time span. Analysis by the project evaluation team showed a pretest–posttest growth that varied by school and grade level, with a mean growth that was 14% above the expected growth over that period. This represented a successful outcome for the initiative and also provided some confidence in the reliability of the instrument.

Following this application of the instrument and to provide some further validation concerning the possibility of consistency in use, the science specialists brought samples of student work for moderation discussions at a program workshop. The purpose was to investigate the teachers' experience with the instrument, their approaches to ensure consistency of judgment within and across year levels, and the potential for developing a consistent language and view of inquiry skills across teachers at the school. Groups discussed and reported on a variety of strategies used to (a) manage the assessment processes and to establish consistency within and across year levels and (b) make reliable judgments about student performance across the class. The discussion included the following dimensions:

Practicalities of running the assessment: Specialists reported being very sensitive to the problem of requiring a significant time commitment by teachers to this task and to the difficulty gathering meaningful data for each student. Several strategies were used to ensure efficient recording of student levels of response, including checklists, the use of iPads by teachers or in one case by students to expedite recording, and the strategic support of teachers using ancillary staff during the session.

Recording strategies: Teachers reported informal strategies for making assessment judgments, including at times a need to make judgments at the group level if they did not have direct access to student discussions as they made decisions, and also the advantage of knowing students well, which they felt enabled them to interpret levels of response to the task.

Developing consistency of judgment: Discussion within groups indicated that, with appropriate negotiation of interpretations of responses, it was possible to achieve agreement on assessment levels. Specialists reported moderation processes within year levels at their schools, but there was less confidence concerning consistency across levels. Specialists reported that the rubric wording developed specific to the tasks was very useful for achieving consistency.

Developing a culture of inquiry teaching and learning: Specialists reported difficulty in having teachers accept that these tasks should be seen as introductions to more deliberate inquiry approaches in teaching science in primary schools. We interpret this as indicating the degree of difficulty in achieving cultural shifts in schools in relation to science curriculum practice.

11.4 Discussion

The experience of the ASELL-SVN and SIA teams in working through these inquiry skill assessments with teachers provided clear evidence of the need for teacher development around understanding and assessing inquiry skills. It was apparent during the workshops with secondary school science teachers that they had little experience using practical work to support skill development and little or no conception of targeting particular skills through structured activities. Similarly, the primary science specialists viewed the SIA as important and interesting, but found the strong and explicit focus on particular inquiry skills challenging; this was especially the case for teachers without a strong science background.

In both projects, we developed a view that for inquiry skill development it was important that an activity focus explicitly on a small number of skills. Often, in practical or investigative work, rubrics deal with all skills in a post hoc assessment. However, if teachers are to genuinely provide targeted support for students in these skills, the design of activities and support structures that focus on one skill—but no more than three—provides the needed clarity. We argue that the development of inquiry skills requires a more deliberate and targeted program than the happenstance approaches that currently prevail. Teachers need support in creating activities, or modifying existing activities, to target specific skills in this way. They need support in developing a language around these skills in order to clearly articulate what the skill entails in different situations and to provide targeted scaffolds to groups of students or the whole class in preparing for or engaging with practical activity.

While it proved difficult in the SIA project to clearly articulate language for the skills that was distinct at half-yearly progression points, in both cases the teams felt that progression in these skills was able to be articulated in a coherent way and that the curriculum framework was effective in supporting this. However, it was also clear that interpretation of the general language for particular context demanded considerable judgment and that teachers would need support to generate these judgments. The problem here was the ambiguity of interpreting particular words in specific situations. Similarly, in the Inquiry Scaffold Tool, the progression in investigative design provision was able to be articulated in a way that made sense to teachers. Here, however, the difficulty was how to describe the type of scaffold that could be provided to support students at the confirmation or structured inquiry level since these supports could be provided in worksheet instructions, in preparatory class discussions, or in other ways.

Within the ASELL-SVN project, teachers were encouraged and trained to take practical activities and modify them to emphasize particular skills. The SIA activities are examples of such modifications, focusing on three skills for each activity. We believe it would be important and very useful, however, to develop resources that provide exemplars of such targeted investigative activities for all levels of the curriculum. In particular, we need to find ways to develop a language of inquiry skill development and assessment that is scalable to the system level. It was clear, in the case of the SIA validation discussions, that teachers do not readily see how these assessments might translate naturally into ways of conceptualizing practical work more generally as skill development. A further challenge is how to develop processes by which teachers can track progression of these skills over a year.

It was also clear, in developing these inquiry skill activities and interpretations, that the skills cannot be seen as independent of conceptual knowledge. The conduct of investigations both depends on and leads to new knowledge. This was definitely the case when considering these activities as representing relations between ideas and evidence. Primary science specialists in the SIA validation workshop pointed out that teachers had seen a strong link between inquiry skills and mathematical processes of data representation; therefore, inquiry skills should not be seen as independent of knowledge creation in the STEM subjects. This is the basis, in fact, of an Australian Research Council project we are currently undertaking to explore the interdisciplinary blending of science and mathematics (https://imslearning.org/). A strong strand of the program is the development of mathematics associated with data modeling, including concepts of variation and sampling. In an ecology unit, for instance, students were supported to develop approaches to counting living things in quadrats and constructing representations—tallies, tables and graphs, and diagrams—comparing counts across different habitats. Thus, the learning of scientific epistemic practices of sampling and dealing with variation is intimately bound up with mathematical concepts of chance and variation, samples, and population.

Associated with this, we have been working on guided inquiry approaches in which students are challenged to visually represent explanations as part of interpreting their inquiry findings. This principle is one of those driving the development of ASELL-SVN activities. Too often, there is an empiricist underpinning of experimentation that assumes results speak for themselves in variable control experiments. For instance, in one investigation of the dissolving of food dye in hot compared to cold water involving Grade 6 (12-year-old children), the teacher emphasized the need to generate a representation as an explanatory model rather than the usual reporting of findings with cursory interpretation. The teacher wrote on the board, during discussion

Explain the spread of the food coloring.

Explain how temperature affected your observations.

Represent your explanations using diagrams or drawings.

Remember: Use the information we have discovered about solids, liquids, and gases.

Figure 11.3 shows a part of one student's work, interpreting the results at the sub-micro level.

Thus, we argue that in focusing strategically on the development of inquiry skills, attention needs to be paid to the link between experimental findings and explanatory modeling so as reflect core epistemic processes of science. It is not enough to simply report results and treat inquiry skills as procedural competencies.

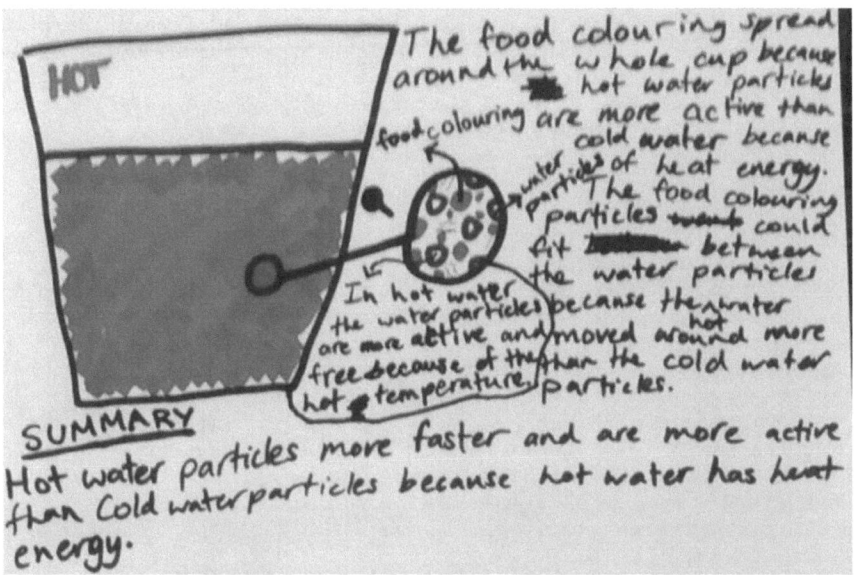

Fig. 11.3 Part of a student's poster explanation of the pattern of dissolving food dye in hot water compared to cold

Acknowledgements ASELL for Schools—Victorian Node academic team included Leader Associate Professor Kieran Lim and Dr. John Long along with Dr. Peta White. The ASELL for Schools project was funded under the Australian Maths and Science Partnership Program. Science Inquiry Assessment team included Professor Russell Tytler, Dr. Gail Chittleborough, and Dr. Peta White. The development and validation process was funded by the Victorian Department of Education and Training.

References

Anderson, R. (2002). Reforming science teaching: What research says about inquiry. *Journal of Science Teacher Education, 13*(1), 1–12.

Australian Assessment, Curriculum and Reporting Authority. (2013). *Australian curriculum: Science*. Sydney, Australia: Author. Retrieved from http://www.australiancurriculum.edu.au/science/content-structure.

Carey, S., Evans, R., Honda, M., Jay, E., & Unger, C. (1989). 'An experiment is when you try it and see if it works': A study of grade 7 students' understanding of the construction of scientific knowledge. *International Journal of Science Education, 11*(5), 514–529.

Chen, H.-L. S., & Tytler, R. (2017). Inquiry teaching and learning: Forms, approaches, and embedded views within and across cultures. In M. Hackling, J. Ramseger, & H.-L. S. Chen (Eds.), *Quality teaching in primary science education: Cross-cultural perspectives* (pp. 93–122). Dordrecht, The Netherlands: Springer. https://doi.org/10.1007/978-3-319-44383-6_5.

Deakin University. (n.d.). Inquiry scaffold tool: Victorian curriculum [Weblog]. Retrieved from https://blogs.deakin.edu.au/asell-for-schools-vic/wp-content/uploads/sites/160/2018/03/Inquiry-Scaffold-Tool-VIC-FINAL.pdf.

Dewey, J. (1996). Essays. In L. Hickman (Ed.), *Collected work of John Dewey, 1882–1953*: The electronic edition. Charlottesville, VA: InteLex.

Duschl, R. (2008). Science education in three-part harmony: Balancing conceptual, epistemic, and social learning goals. *Review of Research in Education, 32*(1), 268–291.

Edwards, D., & Mercer, N. (1987). *Common knowledge: The development of understanding in the classroom*. London, England: Methuen.

Gooding, D. (2004). Visualization, inference and explanation in the sciences. In G. Malcolm (Ed.), *Studies in Multidisciplinarity* (Vol. 2, pp. 1–25)., Multidisciplinary approaches to visual representations and interpretations Amsterdam, The Netherlands: Elsevier.

Goodrum, D., Hackling, M., & Rennie, L. (2001). *Research report: The status and quality of teaching and learning of science in Australian schools*. Canberra, Australia: Department of Education, Training and Youth Affairs. Retrieved from http://www.dest.gov.au/sectors/school_education/publications_resources/profiles/status_and_quality_of_science_schools.htm.

Hassard, J., & Dias, M. (2008). *The art of teaching science: Inquiry and innovations in middle school and high school*. New York, NY: Routledge.

Holmes, N. G., & Wieman, C. E. (2016). Examining and contrasting the cognitive activities engaged in undergraduate research experiences and lab courses. *Physical Review Physics Education Research, 12*(2), 020103.

Latour, B. (1999). *Pandora's hope: Essays on the reality of science studies*. Cambridge, MA: Harvard University Press.

Lehrer, R., & Schauble, L. (2012). Seeding evolutionary thinking by engaging children in modeling its foundations. *Science Education, 96*(4), 701–724.

Masters, G. N. (2015, Feb 26). Learning assessments—Designing the future. *Teacher*. Retrieved from https://www.teachermagazine.com.au/columnists/geoff-masters/learning-assessments-designing-the-future.

Ministry of Education. (2013). *Curriculum guidelines for compulsory education (Grade 1–9): The learning areas of science and technology*. Taipei, Taiwan: Author. Retrieved from http://teach.eje.edu.tw/data/files/class_rules/nature.pdf.

National Institutes of Health. (2005). *Doing science: The process of scientific inquiry*. Bethesda, MD: Author.

National Research Council. (2000). *Inquiry and the national science education standards: A guide for teaching and learning*. Washington, DC: National Academies Press.

National Research Council. (2013). *Next generation science standards: For states, by states*. Washington, DC: National Academies Press. Retrieved from https://www.nap.edu/catalog/18290/next-generation-science-standards-for-states-by-states.

Osborne, J. (2006). Towards a science education for all: The role of ideas, evidence and argument. *Proceedings of the ACER conference: Boosting Science Learning—What will it take?* Camberwell, Australia: ACER.

Palm, T. (2008). Performance assessment and authentic assessment: A conceptual analysis of the literature. *Practical Assessment, Research & Evaluation, 13*(4), 1–11.

Schwab, J. J. (1962). The teaching of science as enquiry. In J. J. Schwab & P. F. Brandwein (Eds.), *The teaching of science* (pp. 3–103). Cambridge, MA: Harvard University Press.

Schwab, J. J. (1964). Structure of the disciplines: Meanings and significances. In G. W. Ford & L. Pugno (Eds.), *The structure of knowledge and the curriculum* (pp. 1–30). Chicago, IL: Rand McNally.

Simon, S., Erduran, S., & Osborne, J. (2006). Learning to teach argumentation: Research and development in the science classroom. *International Journal of Science Education, 28*(2/3), 235–260.

Tytler, R., Prain, V., Hubber, P., & Waldrip, B. (Eds.). (2013). *Constructing representations to learn in science*. Rotterdam, The Netherlands: Sense.

Victorian Curriculum and Assessment Authority. (2015). *Victorian curriculum: Science*. Melbourne, Australia: Author. Retrieved from http://victoriancurriculum.vcaa.vic.edu.au/science/curriculum/f-10

Chapter 12
Assessment Challenges in STEM Reforms and Innovations

Su-Chi Fang and Ying-Shao Hsu

12.1 Introduction

STEM education, by design, crosses disciplinary boundaries. It has been interpreted along an integration continuum as a collection of core concepts and skills from isolated disciplines (multidisciplinary), linked concepts and skills from two or more disciplines (interdisciplinary), or application of concepts or skills from more than two disciplines to real-life problems or projects (transdisciplinary) (Vasquez, Sneider, & Comer, 2013). However, there seems to be no agreement on the definition of STEM. The term *integrated STEM* is intentionally and specifically used for interpreting STEM education as teaching and learning STEM knowledge and practices in a more connected manner and within the context of real-world issues (National Academy of Engineering & National Research Council [NRC], 2014). More specifically, the intention of integrated STEM education is to provide authentic learning contexts where students can learn, consolidate, and apply disciplinary knowledge and skills in an integrated manner through solving real-world problems and creating engineering solutions. Assessments of STEM learning, therefore, need to align with this specific intention and to construct trustworthy indicators of students' STEM learning outcomes.

In contrast to disciplinary-based learning, integrated STEM teaching and learning purposely offer ample opportunities for the development of competencies and higher order thinking skills other than simply learning isolated subject content knowledge. Thus, to assess STEM learning effectively as well as adequately, teachers and educators must develop an inclusive assessment system that allows students to demonstrate their ability to apply integrated knowledge in different contexts, to solve problems successfully, or to create adequate solutions through various types of assessments.

Ying-Shao Hsu is a visiting professor at University of Johannesburg, South Africa.

S.-C. Fang (✉) · Y.-S. Hsu
Graduate Institute of Science Education, National Taiwan Normal University, Taipei, Taiwan
e-mail: suchhii@msn.com

© Springer Nature Singapore Pte Ltd. 2019
Y.-S. Hsu and Y.-F. Yeh (eds.), *Asia-Pacific STEM Teaching Practices*,
https://doi.org/10.1007/978-981-15-0768-7_12

In other words, assessment for integrated learning needs to consider not only various aspects of learning outcomes but also include various and valid assessment types or formats that can truly reflect what students have learned in an integrated STEM environment.

In addition to considering *what* is to be examined in assessments, it is crucial to contemplate *when* in terms of the delivery points for assessment during STEM instruction. Depending on the time of administration and the purposes of the data, assessments can be classified into three types: diagnostic, formative, and summative. Diagnostic assessments may be used for appraising students' prior knowledge and skills before instruction, and these data can inform the teacher's lesson planning and differentiated instruction. The purpose of formative assessments is to evaluate what students have learned after a short period of time. These data can inform learners' actions and help the teacher document students' learning status, provide remediation for unrealized outcomes, and adjust future teaching. Summative assessments are normally administered at the end of different levels of learning processes, such as a chapter, a semester, or a grade level. An inclusive assessment system for STEM needs to encompass these three types of assessments so that students' learning progression can be captured consistently and comparably.

Despite that STEM education has been a major movement in the past decade (Martín-Páez, Aguilera, Perales-Palacios, & Vílchez-González, 2019), research efforts and teacher preparation and professional programs appear to focus more on STEM instructional design. Assessment approaches and their development are nearly overlooked (Sondergeld, Koskey, Stone, & Peters-Burton, 2015). Thus, this chapter is intended to initiate and facilitate conversations and discussions on the assessment issues related to integrated STEM teaching and learning. First, we reviewed the STEM education research database used in Chap. 1 (Hsu & Fang, this book) in an attempt to understand current assessment issues and approaches for STEM learning. Second, we employed the idea of the *assessment triangle* proposed by NRC (2001) as a critical lens to further examine the current assessments for STEM learning and to identify significant assessment issues in STEM education. Third, we extended the notions of *what* and *when* in assessments to a broader scale and used these insights as a foundation for a multilevel-multifaceted STEM assessment framework.

12.2 Current Assessments for STEM Learning: Perspectives from a Review of STEM Research

The exploration of current assessments used for STEM learning was confined to the database of studies in Chap. 1 that had adopted assessment(s) or tool(s) to examine student learning achievement (12 articles). Three major types of assessment were identified from the review: subject-content-knowledge-focused tests (2 articles), standardized tests (6 articles), and project-specific assessments (4 articles).

12.2.1   Subject-Content-Knowledge-Focused Assessment Tests

The subject-content-knowledge-focused tests are used for measuring students' understanding of a particular concept. Korur, Efe, Erdogan, and Tunç (2017) used the simple machine achievement test; this test includes true/false, matching, and multiple-choice items to investigate the effects of different instructional approaches (e.g., teacher-directed instruction and scaffolded-design-based learning) on students' understanding of the concepts of a simple machine after the STEM learning experience (i.e., design a toy crane). Similarly, Schnittka, Evans, Won, and Drape (2016) adopted the heat and transfer evaluation, a 12-multiple-choice-item test, to determine the students' concept attainment after the engineering-design-based, after-school learning on the topic Save the Penguins.

12.2.2   Standardized Tests

The standardized measures that focused on content knowledge in separate disciplines were prevalent in large-scale STEM studies, such as grade point average (GPA), American College Testing (ACT), and statewide test scores. These assessment results were normally obtained from the government and were used for indicating longitudinal impacts of a STEM program on students' academic achievements. Han, Capraro, and Capraro (2015) explored how students' demographic backgrounds and performance proficiency levels affected students' mathematics achievements after a 3-year STEM-PBL (i.e., project-based learning) intervention. They used the mathematics scores from the Texas Assessment of Knowledge and Skills as their outcome variable. Dickerson, Eckhoff, Stewart, Chappell, and Hath-cock (2014) compared the state-standardized test scores on science, mathematics, and English reading for students who attended a pullout STEM program with those who did not attend the STEM program. Likewise, Micari, Van Winkle, and Pazos (2016), Means, Wang, Young, Peters, and Lynch (2016), and Means et al. (2017) adopted GPA and/or ACT scores to evaluate what extent the students' learning experiences of a university Gateway Science Workshop program or inclusive STEM high schools influenced their academic learning outcomes. Unlike the above studies that simply used standardized tests to measure students' academic learning outcomes, Lamb, Akmal, and Petrie (2015) considered how a whole-school integrated STEM program might impact integrated constructs in addition to the content knowledge. They identified two integrated constructs—spatial visualization and mental rotation—and adopted the paper-folding test and Shepard and Metzler Mental Rotation Test to evaluate students' cognitive development after the integrated STEM program.

12.2.3   Project-Specific Assessment Tests

Project-specific assessments were usually developed by researchers or teachers to evaluate student performance and artifacts during or after a particular STEM program. Two of the four project-specific assessment studies were concerned with students' in-process performance, so they developed and used the tests as formative evaluations. King and English (2016) involved students in an optical engineering design activity. They analyzed students' workbook design sketches to understand how students applied STEM concepts in a design process. A five-level coding scheme developed from the empirical data indicated students' various degrees of sophistication and accuracy in applying core science and mathematics concepts in the design process. Chien and Chu (2017) required students to complete a worksheet at each learning phase during an 8-week STEAM engineering program on CO_2-car design. The worksheet consisted of questions for evaluating students' STEAM knowledge, data collection, design drafts, group discussion records, forest outcomes, race outcomes, correction-and-improvement discussion records, and opinions and feedback on the course.

Guzey, Ring-Whalen, Harwell, and Peralta (2019) developed science and engineering achievement tests based on some public item banks: TIMSS, National Assessment of Educational Progress (NAEP), and American Association for the Advancement of Science. The tests were used for examining students' yearly learning outcomes and how they are related to different engineering integration approaches (e.g., add-on, implicit, explicit) during the three design-focused life science units. Sahin, Gulacar, and Stuessy (2015) adopted an online self-report survey that required students to indicate the degree to which their participation in an International Sustainable World Energy, Engineering, and Environmental project supported them to improve their abilities in 10 selected twenty-first-century skills.

In summary, the review showed that the types of assessments are diverse in these STEM studies but that most were focused on content knowledge in separate disciplines. Few studies assessed and discussed how students' inquiry abilities, higher order thinking skills, or creativity could be improved in STEM learning. The subject-content-knowledge-focused tests had relatively narrow content foci whereas the standardized tests were likely to have items that only partly aligned with an integrated STEM curriculum or program. The project-specific assessments, on the other hand, were inclined to evaluate very specific learning performances during a particular STEM learning experience. Therefore, the application of these tests was limited in specific STEM learning contexts. It is worth noting that only one (Lamb et al., 2015) of the 12 studies reviewed measured spatial visualization and mental rotation as integrated constructs and attempted to explore STEM learning in a more integrated manner. We would argue that an integrated STEM curriculum by design requires students to learn the four disciplines in a more connected, holistic way. Therefore, integrated STEM assessments should identify the set of knowledge—integrated or separate—and core competencies to be learned, developed, and applied during learning activities, which can be monitored during and tested at the end of the program.

12.3   Issues in Assessing Students' Learning Outcomes Through STEM

The reviewing results presented in the previous section is based on the analysis of *what* and *when* viewpoints. The assessments in the 12 studies were delivered across different levels (when), but most were focused on disciplinary knowledge instead of the expected learning emphasized in integrated STEM—such as inquiry abilities, problem-solving abilities, higher order thinking skills, or other competencies (what). As mentioned earlier, assessment is one of the significant features in the educational system and indeed assessments are not separate from but entangled with standards, teaching, and learning. Quality assessments have to align with standards, school curriculum, and classroom instruction; the design and development of assessments should correspond with standards and instructional goals.

12.3.1   The Alignment Between Assessment and Integrated STEM Curriculum

Knowing What Students' Know: The Science and Design of Educational Assessment (NRC, 2001) proposed the assessment triangle (i.e., cognition, observation, interpretation) that describes assessment as a process of reasoning from evidence. Specifically, the cognition corner of the triangle implies the set of essential knowledge and skills being identified in the learning process with the lens provided by a particular theory of learning. The observation corner represents the assessment tasks through which students can perform their understandings and skills corresponding to those identified in the cognitive model. The interpretation corner refers to the selection and application of the methods and tools being applied in order to reason from the data obtained in the observations. Importantly, the three elements in the assessment triangle are connected and interdependent—they must be "in synchrony" (NRC, 2001, p. 44). The notion of synchrony highlighted in the assessment triangle provides a more critical and holistic view to examine our review of current assessments for STEM learning.

As shown in Chap. 1 (Hsu & Fang, this book), our review indicated that some studies did not specify the goals of the STEM programs, nor did they express how their curricula and instruction were informed and transformed based on their goals. The lack of well-defined goals may bring serious consequences to assessment development—If you do not know where you want to go, then any direction will suffice. It is worth noting that, without clear goals, the interpretation of assessment data in terms of the impact of the STEM program on certain aspects of student learning performance might also be problematic and misleading. Moreover, the learning objectives in the reviewed studies were not clearly defined; most of the STEM programs set their goals in a very general manner, for example, to maintain and increase students' interest in science and technology or to learn about STEM content and careers. More specific learning objectives such as the intended knowledge, practices, or skills (the cognition corner) to be learned and developed were not explicitly described. Although the studies explained both the purpose and the procedures to use the assessment tools, how the assessments are aligned with and linked to the goals were often overlooked. In other words, the cognition, observation, and interpretation aspects of the assessment triangle in the current assessments for STEM reports were not clearly defined, nor were they coherently connected or synchronized.

12.3.2 Assessments for Integrated Learning

As stated in the introduction of this chapter, it is commonly accepted that STEM learning values interdisciplinary or transdisciplinary views of knowledge and practices. However, the current assessments for STEM learning in the reviewed research tended to purely focus on the evaluation of isolated content knowledge in separate disciplines, which is a multidisciplinary view of STEM. The tasks used in the current assessments did not align with the integrated learning experience in which the individual STEM disciplinary contents are integrated into real-world problems or engineering design process interlaced with open-ended inquiry. Assessment needs to align with interdisciplinary, transdisciplinary, or integrated views of STEM teaching and learning; therefore, we need to design and develop assessment tasks that require students to apply integrated knowledge, engineering, and scientific practices and to integrate inquiry, design, and reasoning so as to solve real-world problems or produce innovative solutions. That is to say, it is crucial to reconsider the observation corner of the assessment triangle in terms of the types of data that can be counted as evidence in integrated STEM learning.

One significant question that should be raised and contemplated in STEM education assessment is what needs to be assessed, when does it need to be done, and how should these data be collected. The 12 studies reviewed demonstrated that STEM curricula are highly diverse. Integrated STEM is often situated in real-world contexts, where different contexts involve different sets of content knowledge and practices and require application of various higher order thinking skills. However, educators and researchers have not reached consensus about what competencies or concepts need to be developed via integrated STEM. Moreover, individual differences in student-centered learning environments may further complicate the problem about how to assess students' progress on these focused competencies and conceptual understandings. Some curriculum reform documents have started to address this issue; A Framework for K–12 Science Education (NRC, 2012) proposed bringing engineering practices into inquiry as a way to teach science, and the Next Generation Science Standards (NRC, 2013) developed a set of standards for integrating engineering into science learning across grade levels. These standards suggest possible competencies and a knowledge hierarchy for STEM learning. However, more research is needed to identify which are core STEM competencies and how to best arrange them in a cohesive, coherent manner.

12.4   Measuring the Impacts of Integrated STEM Curricula: The Idea of a Multilevel-Multifaceted Approach

The studies reviewed in this chapter involve various integrated STEM curricula being implemented at different scales: single learning unit, multiple learning units, whole-school program, and inclusive STEM high schools. As the implementation scales of the integrated STEM curricula were different, the assessments applied to the evaluation of the curricula varied in its nature and purpose. The different scales of curricula can be seen as a learning continuum, and the assessments led us to contemplate how learning transfer can be assessed and monitored in STEM learning. That is, how the impacts of integrated STEM learning can be measured across a single unit of study, several units of study, a whole year of schooling, or a total school program.

Ruiz-Primo, Shavelson, Hamilton, and Klein (2002) pointed out that measuring the impacts of educational reform is challenging. This is because the reform goals and standards are likely to be translated into diverse programs and practices across different levels of the educational system from states to districts and schools. We contend that the situation in integrated STEM education is actually similar to that in educational reform. As shown in the reviewed studies and elsewhere, integrated STEM curricula take various forms and scales across different schools, districts, and states. Furthermore, similar situations pose major challenges to evaluation as the impact of integrated STEM curricula/educational reform is presumably variable across settings. Ruiz-Primo et al. (2002) proposed a multilevel-multifaceted approach to address the challenge of evaluating educational reform, and we suggest that this approach is potentially useful for measuring the impacts of integrated STEM learning.

A multilevel-multifaceted approach consists of two notions. The first notion is that learning achievement is multifaceted; thus, different types of assessments are required to attend to different facets of knowledge and skills. The second notion concerns multiple levels of assessment varying from the closest to the distant point in relation to the enactment of the curriculum (i.e., immediate, close, proximal, distal, remote). The design of integrated STEM assessments, therefore, needs to include various types of tests or data collection techniques addressing students' development of disciplinary or integrated knowledge, competencies, literacy, and attitudes at multiple levels of proximity, as shown in Fig. 12.1.

Well-integrated instruction provides opportunities for students to apply and connect STEM concepts, enhance their use of higher level thinking skills (e.g., problem-solving abilities and computational thinking skills), and cultivate their attitude toward STEM-related professions. For example, a knowledge integration framework composed of four phases—eliciting ideas, adding ideas, distinguishing ideas, and sorting out ideas—provides guidelines to help students develop and connect concepts across different disciplines (Linn & Eylon, 2011). Other instructional approaches (e.g., problem-based learning and engineering design) embedded in STEM lessons provide students with experiences of how to integrate and apply concepts across

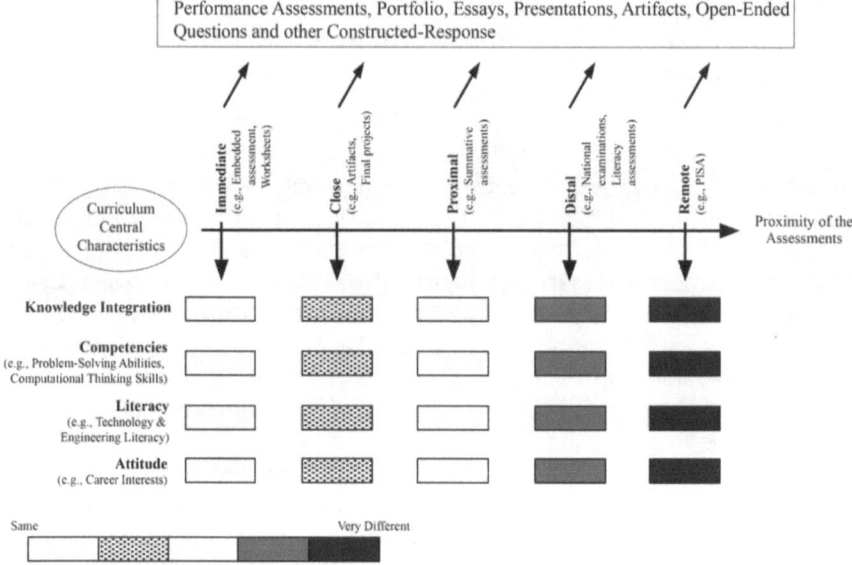

Fig. 12.1 The multilevel-multifaceted STEM assessment framework

disciplines in real-world situations (Lou, Shih, Ray Diez, & Tseng, 2011). What students have learned in these integrated learning environments might be very different from that in traditional, disciplinary-based learning environments. Thus, the various facets of STEM learning need to be identified and included in the assessments. In our assessment framework for STEM learning (Fig. 12.1), the facets of STEM learning we propose correspond to different dimensions of learning outcomes that occurred during learning activities—including knowledge integration, competencies (e.g., problem-solving ability and computational thinking skills), literacies (e.g., technology and engineering literacy), and attitude (e.g., career interests and motivation)—need to be incorporated.

The dimension of multiple levels of proximity is associated with assessment delivery points. The assessments are administered at several levels to monitor students' learning and to indicate how far the STEM learning outcomes transfer. Different levels of assessment administration also imply varied intentions and purposes and serve different functions. At the *immediate level*, teachers use embedded assessments such as worksheets to track students' learning. Thus, instant feedback can be provided to address students' needs during the learning process. The purposes of immediate assessments, therefore, are to diagnose students' learning and to inform the following instruction. At the *close level*, teachers may need to develop scoring rubrics to evaluate student performance or to appraise the project reports or artifacts created during STEM learning activities. At the *proximal level*, teachers may give students different STEM problems to solve in order to examine how well they are able to transfer what they have learned. At the *distal level*, student learning performances are assessed based on the national standards. At the *remote level*, assessments are conducted via general measurements of international comparison tests such as PISA. Indeed, the functions of the assessments are not determined by the levels they are but depend on how the data are used. Assessment results can be summative for the currently finished learning stage (assessment of learning); however, at the same time, it can be seen as a formative report for a long-term learning process (assessment for learning).

STEM-related competencies are often recognized as problem-solving abilities (English, Hudson, & Dawes, 2013; English, King, & Smeed, 2017) and computational thinking skills (Sengupta, Dickes, & Farris, 2018; Weese & Feldhausen, 2017). English et al. (2017) utilized engineering design to guide students to learn design-based problem-solving; they used an open-coding technique to interpret students' activity booklets and identify the features of problem-solving (e.g., designing and constructing, assessing design, and redesigning and reconstructing). According to the idea of *assessment as learning* proposed by Earl (2013), if teacher evaluation criteria are provided during learning or students generate their own evaluating criteria, students can monitor what they are learning personally or in groups and reflect on their observations to make modifications, adjustments, or major changes in what they understand or create.

One commonly used approach for measuring computational thinking skills is evidence-centered assessment (Weese & Feldhausen, 2017) that provides a real-time tool for teachers to obtain information about what students are struggling with or how they are succeeding during STEM learning (Basawapatna, Repenning, & Koh, 2015). Based on that information, teachers can diagnose students' learning difficulties and identify students' learning patterns of computational thinking so that they can inform, adjust, or redesign the ways they teach. Some researchers develop self-efficacy surveys to detect students' belief that they can accomplish a STEM task as evidence of their computational thinking ability (Weese & Feldhausen, 2017).

It is assumed that when students participated in STEM learning activities they had more chances to increase their exposure to STEM knowledge and then enhance their career interests in STEM-related jobs (Lou et al., 2011). Some studies have looked into how STEM education promotes students' career interests and motivation (Guzey, Harwell, & Moore, 2014; Tseng, Chang, Lou, & Chen, 2013). A survey was a common approach to assess students' attitude toward STEM, for example, Guzey et al. (2014) developed a survey instrument that measured student attitude toward STEM learning and STEM career.

Technological and engineering literacy (TEL) is broadly defined as the capacity to use, understand, and evaluate technology as well as to understand technological principles and strategies needed to develop solutions and achieve goals.

The TEL assessment is designed to measure three interconnected areas of technology and engineering experience in and out of the classroom: technology and society, design and systems, and information and communication technology (National Assessment Governing Board, 2013). Exemplar assessment items for these goal areas can be found under technological and engineering literacy on the NAEP website (https://www.nationsreportcard.gov/tel/about/assessment-framework-design/).

12.5   An Example of Enacting the Multilevel-Multifaceted STEM Assessment Framework

This section provides an example of the multilevel-multifaceted STEM assessment framework regarding advancing integrated STEM learning via engineering design (English et al., 2017). English and her colleagues applied a five-step process of engineering design to guide students' learning of design-based problem-solving including problem scoping (understand problem boundaries), idea creation (formulate ideas and develop a plan), designing and constructing (sketch design, interpret design, transform to models), assessing design (check constraints and test models), and redesigning and reconstructing (sketch second design and transform to models). They selected earthquakes as a problem to engage sixth-grade students in constructing a building to withstand earthquake damage using toothpicks and plasticine. English et al. used a technique of open coding to interpret students' activity booklets and identify the features of students' sketches, such as meeting the constraints, accurately labeling materials, displaying engineering techniques, indicating measurements, costs and material qualities, and accurately labeling shapes or features of the structures. These sketching features can be recognized as abilities within design-based problem-solving, such as designing and constructing (e.g., accurately labeling materials, displaying engineering techniques, accurately labeling shapes or features of the structures) and assessing design (e.g., meeting the constraints; indicating measurements, costs, and material qualities). Furthermore, they examined how students applied their STEM disciplinary knowledge using an interview protocol that asked students about (a) their learning from the first and second building tests and (b) the redesign changes they would make and why. The evidence of students' learning STEM disciplinary knowledge related to the earthquakes problem included concepts such as the base of the building, weight, complexity of the design, material use, balance, stability, strength of the structure, and mathematics use for measurements.

This example of assessing artifacts and student responses demonstrated how a multifaceted STEM assessment that included design-based, problem-solving abilities and STEM disciplinary knowledge can be developed during integrated STEM learning. The addition of a distal assessment of design-based problem-solving abilities to detect how students transfer their problem-solving abilities to a different problem such as building a safe bridge or a solar automatic trolley (Tseng et al., 2013) makes a multilevel STEM assessment applicable. The multilevel-multifaceted STEM assessment framework (Fig. 12.1) can serve as guidance for STEM educators and researchers to examine students' STEM learning performance and transfer during their continuity of integrated STEM education.

12.6 Conclusion

This review of 12 research studies revealed that there are varied types of assessment adopted for evaluating student learning outcomes at different scales for a single learning unit, multiple learning units, a whole-school program, and an inclusive STEM high school. The findings show that students can benefit from integrated STEM learning experiences in many ways. Nevertheless, it is noted that most of the assessments were focused on disciplinary knowledge. More research is needed for developing competency-based assessments to investigate how and to what extent STEM learning can cultivate students' development of inquiry abilities, higher order thinking skills, or creativity. The assessment triangle underscores a coherent, cohesive linkage between the intended knowledge and skills (cognition), the assessment tasks (observation), and the analysis and interpretation of assessment results (interpretation). Therefore, it is important to, first, identify the intended STEM learning outcomes (knowledge, attitudes, or skills) to be developed in the learning process and, second, contemplate how to design assessments that really create evidence of STEM learning. The multilevel-multifaceted STEM assessment framework proposed in the chapter provides guidance on the design and development of assessments for determining various facets of learning outcomes at different levels of proximity.

Acknowledgements This work was financially supported by the Institute for Research Excellence in Learning Sciences of National Taiwan Normal University from the Featured Areas Research Center Program within the framework of the Higher Education Sprout Project and Ministry of Science and Technology 107-2511-H-003-043-MY3 Project by the Ministry of Education in Taiwan.

References

References marked with an asterisk indicate the 12 studies included in the review

Basawapatna, A., Repenning, A., & Koh, K. H. (2015). Closing the cyberlearning loop. In *Paper Presented at the 46th ACM Technical Symposium on Computer Science Education—SIGCSE'15*, Kansas City, MO, USA.
*Chien, Y. H., & Chu, P. Y. (2017). The different learning outcomes of high school and college students on a 3D-Printing STEAM engineering design curriculum. *International Journal of Science and Mathematics Education, 16*(6), 1047–1064.
*Dickerson, D. L., Eckhoff, A., Stewart, C. O., Chappell, S., & Hathcock, S. (2014). The examination of a pullout STEM program for urban upper elementary students. *Research in Science Education, 44*(3), 483–506.
Earl, L. M. (2013). *Assessment as learning: Using classroom assessment to maximize student learning* (2nd ed.). Thousand Oaks, CA: Corwin/Sage.
English, L. D., Hudson, P. B., & Dawes, L. A. (2013). Engineering based problem solving in the middle school: Design and construction with simple machines. *Journal of Pre-College Engineering Education Research, 3*(2), 1–13.

English, L. D., King, D., & Smeed, J. (2017). Advancing integrated STEM learning through engineering design: Sixth-grade students' design and construction of earthquake resistant buildings. *Journal of Educational Research, 110*(3), 255–271.

Guzey, S. S., Harwell, M., & Moore, T. (2014). Development of an instrument to assess attitudes toward science, technology, engineering, and mathematics (STEM). *School Science and Mathematics, 114*(6), 271–279.

*Guzey, S. S., Ring-Whalen, E. A., Harwell, M., & Peralta, Y. (2019). Life STEM: A case study of life science learning through engineering design. *International Journal of Science and Mathematics Education, 17*(1), 23–42.

*Han, S., Capraro, R., & Capraro, M. M. (2015). How science, technology, engineering, and mathematics (STEM) project-based learning (PBL) affects high, middle and low achievers differently: The impact of student factors on achievement. *International Journal of Science and Mathematics Education, 13*(5), 1089–1113.

*King, D., & English, L. D. (2016). Engineering design in the primary school: Applying stem concepts to build an optical instrument. *International Journal of Science Education, 38*(18), 2762–2794.

*Korur, F., Efe, G., Erdogan, F., & Tunç, B. (2017). Effects of toy crane design-based learning on simple machines. *International Journal of Science and Mathematics Education, 15*(2), 251–271.

*Lamb, R., Akmal, T., & Petrie, K. (2015). Development of a cognition-priming model describing learning in a STEM classroom. *Journal of Research in Science Teaching, 52*(3), 410–437.

Larkin, K., & Jorgensen, R. (2018). *STEM Education in the junior secondary: The state of play.* Singapore: Springer.

Linn, M. C., & Eylon, B.-S. (2011). *Science learning and instruction: Taking advantage of technology to promote knowledge integration.* New York, NY: Routledge.

Lou, S.-J., Shih, R.-C., Ray Diez, C., & Tseng, K.-H. (2011). The impact of problem-based learning strategies on STEM knowledge integration and attitudes: An exploratory study among female Taiwanese senior high school students. *International Journal of Technology and Design Education, 21*(2), 195–215.

Martín-Páez, T., Aguilera, D., Perales-Palacios, F. J., & Vílchez-González, J. M. (2019). What are we talking about when we talk about STEM education? A review of literature. *Science Education.* Advance online publication. https://doi.org/10.1002/sce.21522.

*Means, B., Wang, H., Wei, X., Lynch, S., Peters, V., Young, V., et al. (2017). Expanding STEM opportunities through inclusive STEM-focused high schools. *Science Education, 101*(5), 681–715.

*Means, B., Wang, H., Young, V., Peters, V. L., & Lynch, S. J. (2016). STEM-focused high schools as a strategy for enhancing readiness for postsecondary STEM programs. *Journal of Research in Science Teaching, 53*(5), 709–736.

*Micari, M., Van Winkle, Z., & Pazos, P. (2016). Among friends: The role of academic-preparedness diversity in individual performance within a small-group STEM learning environment. *International Journal of Science Education, 38*(12), 1904–1922.

National Academy of Engineering & National Research Council. (2014). *STEM integration in K-12 education: Status, prospects, and an agenda for research.* Washington, DC: National Academies Press.

National Assessment Governing Board. (2013). *2014 abridged technology and engineering literacy framework for the 2014 national assessment of educational progress.* Washington, DC: Author. Retrieved from http://purl.fdlp.gov/GPO/gpo44685.

National Research Council. (2001). *Knowing what students know: The science and design of educational assessment.* Washington, DC: National Academies Press.

National Research Council. (2012). *A framework for K-12 science education: Practices, crosscutting concepts, and core ideas.* Washington, DC: National Academies Press.

National Research Council. (2013). *Next generation science standards: For states, by states.* Washington, DC: National Academies Press.

Rennie, L., Venville, G., & Wallace, J. (2012). *Integrating science, technology, engineering, and mathematics: Issues, reflections, and ways forward*. New York, NY: Routledge/Taylor & Francis.

Ruiz-Primo, M. A., Shavelson, R. J., Hamilton, L., & Klein, S. (2002). On the evaluation of systemic science education reform: Searching for instructional sensitivity. *Journal of Research in Science Teaching, 39*(5), 369–393.

*Sahin, A., Gulacar, O., & Stuessy, C. (2015). High school students' perceptions of the effects of international science olympiad on their STEM career aspirations and twenty-first century skill development. *Research in Science Education, 45*(6), 785–805.

*Schnittka, C. G., Evans, M. A., Won, S. G. L., & Drape, T. A. (2016). After-school spaces: Looking for learning in all the right places. *Research in Science Education, 46*(3), 389–412.

Sengupta, P., Dickes, A., & Farris, A. (2018). Toward a phenomenology of computational thinking in STEM education. In M. S. Khine (Ed.), *Computational thinking in the STEM disciplines: Foundations and research highlights* (pp. 49–72). Cham, CH: Springer.

Sondergeld, T. A., Koskey, K. L. K., Stone, G. E., & Peters-Burton, E. E. (2015). Data-driven STEM assessment. In C. C. Johnson, E. E. Peters-Burton, & T. J. Moore (Eds.), *STEM road map: A framework for integrated STEM education*. New York, NY: Routledge.

Tseng, K.-H., Chang, C.-C., Lou, S.-J., & Chen, W.-P. (2013). Attitudes towards science, technology, engineering and mathematics (STEM) in a project-based learning (PjBL) environment. *International Journal of Technology and Design Education, 23*(1), 87–102.

Vasquez, J. A., Sneider, C., & Comer, M. (2013). *STEM lesson essentials, grades 3–8: Integrating science, technology, engineering, and mathematics*. Portsmouth, NH: Heinemann.

Weese, J. L., & Feldhausen, R. (2017, June). STEM outreach: Assessing computational thinking and problem solving. In *Paper presented at the 124th American Society for Engineering Education Annual Conference and Exposition (ASEE 2017)*, Columbus, OH, USA.

Chapter 13
Epilogue—Understanding STEM for STEM Education: Toward a Systems Approach

Sibel Erduran

13.1 Introduction

In recent years, there have been numerous calls and initiatives to highlight the importance of science, technology, engineering, and mathematics (STEM) and its teaching in secondary education. As some of the chapters in this book also highlight, many international curricula (e.g., Chap. 5) and research studies (e.g., Chap. 10) have taken STEM education as a focal point of advocacy and investigation. This book provides a comprehensive overview of some of the key trends in international research and development efforts in STEM education in different national contexts (e.g., Taiwan, Korea, Australia). The primary purpose of this final chapter is twofold: (a) to review some of the key themes raised in the book in order to provide a summary of the main themes covered in the 12 chapters, resulting in some qualitative trends in the way that educators are engaging in research on STEM education, and (b) to consider the implications for potential future research and development efforts on STEM education. The chapters are multifaceted in content; hence, any review will be limited by definition. The approach underpinning the chapter is to thematically organize some major issues represented in the chapters in order to provide a summary of the overall content. Issues related to curriculum, domain knowledge, teaching, learning, school culture, and assessment will be highlighted in relation to STEM education. Subsequently, the implications of these issues for future research and development will be considered in anticipation of extending the agenda of the book.

S. Erduran (✉)
University of Oxford, Oxford, UK
e-mail: sibel.erduran@education.ox.ac.uk

© Springer Nature Singapore Pte Ltd. 2019
Y.-S. Hsu and Y.-F. Yeh (eds.), *Asia-Pacific STEM Teaching Practices*,
https://doi.org/10.1007/978-981-15-0768-7_13

13.2   Summary of Key Themes

The introductory chapter provides a fairly extensive account of the background to research on STEM education. Hsu and Fang problematize how and to what extent integrated STEM learning experiences may foster student creativity, support the development of higher order thinking skills, and impact their epistemological beliefs and views about science learning (Chap. 1). A significant contribution of the chapter is a framework that the authors propose for the development of STEM curricula. The framework involves the following steps: (a) identify core competencies, (b) select a real-world context or problem, (c) prepare supporting resources and tools, (d) design a series of activities to engage students, and (e) develop an evaluation rubric for assessing the selected core competencies. The theme of curriculum development is again picked up in Chap. 2 when Chu, Son, Martin, and Treagust contrast integrated STEM provision in Korea and Australia. The latter authors provide examples of how Korea incorporated an arts-integrated STEM orientation while Australia has had no such orientation—although other broad themes such as intercultural understanding, literacy, and sustainability have been advocated in the curriculum. Chu and colleagues illustrate the benefits and challenges of arts-integrated STEM approaches for teachers' classroom practice in both countries.

Two chapters deal with issues related to domain knowledge in STEM. In Chap. 7, Fan and Yu stress the core value of STEM education as providing balanced opportunities of hands-on and minds-on learning. They focus on engineering design, which raises questions about teachers' pedagogical content knowledge. The domain of mathematics is covered in Chap. 8 where Lee, Lim, De, and Hilmy describe the process of learner intuitions and disciplinary intuitions in STEM. To illustrate the process of enacting a STEM curriculum in a Singapore public school, they present an in-depth case study of a collaborative effort to develop a STEM learning framework into the secondary mathematics curriculum of the school dealing with averages. The purpose of this effort was to promote greater authenticity in students' learning of mathematical data handling skills, applications, and interests using a real-world learning context involving localized data collection from open-source sensors and a real-world problem-solving task. The authors found that a STEM learning framework supported by open-source sensors can indeed be used to create an authentic learning environment, which in turn can surface and leverage learner intuitions and enhance learning.

The central role of teachers and teaching in STEM education is acknowledged in numerous chapters. Hsu and Fang in Chap. 1 highlight the significance of educating teachers, both preservice and in-service teachers, in STEM given the relative unfamiliarity of STEM in school subjects. A series of chapters then provide accounts of how teachers' knowledge of STEM and STEM teaching can be improved. In Chap. 3, Chan, Yeh, and Hsu propose a theoretical framework for examining and analyzing teachers' practical knowledge for STEM teaching. Chan and colleagues synthesize literature to identify features of teacher knowledge needed for effective

STEM teaching. On the other hand, Chan and Yip in Chap. 5 report on STEM education in Hong Kong where it became a mandatory part of the science curricula for secondary schools in 2016. The authors observe that curriculum reform poses a great challenge to teachers who were educated to be discipline-specific. They compare an experienced and novice teachers' conceptions about STEM and their practical knowledge for STEM teaching using in-depth interviews. The theme of teaching and teachers is extended in Chaps. 4 and 9 where the broader teacher community aspects are discussed. In Chap. 4, Yeh and Hsu discuss a study conducted in a high school in Taiwan where a professional learning community model was used to develop a series of STEM curricula. They describe the model including its courses, and they score the teachers' instructional knowledge of STEM. Xu, Campbell, and Hobbs in Chap. 9, on the other hand, reflect on teachers from primary schools in Australia who were involved in a professional learning program specifically designed to build their confidence and capacity for teaching STEM through inquiry-based approaches. They draw on data of changing teacher STEM pedagogy to generate insight into the diverse responses that schools can have to professional learning in STEM and entrepreneurial thinking. The findings of the study indicate the importance of research-informed frameworks that are flexible enough to be applied to schools.

The argument for teaching of STEM in an integrated fashion is ultimately based on the need to facilitate students' learning in a meaningful manner and to develop a range of skills such as problem-solving and critical thinking. Two chapters in the book deal with student learning of STEM. In Chap. 6, So, Li, and He explore a fairly under-examined topic of special educational needs (SEN) in relation to STEM education. They investigate special school teachers as part of a STEM education professional development program. They note a positive response from teachers in designing STEM learning for special educational needs. The researchers put forward the recommendations by the teachers for the adaptation of resources and strategies for SEN purposes. For example, the teachers suggested strategies to meet students' needs with smaller steps/procedures, division of tasks among students with different abilities, provision of relevant background knowledge, and being equipped with relevant tools and awareness of safety precautions. Among other strategies, So and colleagues suggest (a) integrating advance technologies (e.g., 3D-printing, robotics, virtual reality, microprocessors) to achieve technology focus in learning and (b) organizing a variety of outside-school STEM activities (e.g., exhibition, company visits) to motivate students to acquire through experience the necessary STEM skills. Furthermore, assessment has always been a challenge to new initiatives in learning and teaching.

A significant challenge in the integration of STEM in teaching and learning can be a lack of coherence with both formative and summative assessments. When the goals of the curriculum switch from the teaching of a particular domain like science to the goal of teaching topics that bring all STEM subjects together, a new vision is needed for assessment. Three chapters in the book deal with issues related to assessment. In Chap. 6, So, Li, and He propose strategies to measure the positive effects of STEM education on students' STEM perception, attitudes, skills, and career aspirations. This framework, though related to assessment, is primarily a research orientation. Two other chapters deal with assessment in the context of STEM education in a more direct sense. In Chap. 11, Tytler and White argue that the focus on STEM as an interdisciplinary construct has placed greater emphasis on the contextual applications of science and mathematics practices. For science, this has resulted in a need to more clearly articulate the nature of scientific epistemic practice, including supporting and assessing scientific inquiry skills. They describe a project that involved the development and trialing of secondary level practical inquiry activities representing contemporary STEM practices and ideas. Teachers and students engaging with these activities were introduced to an inquiry scaffold tool that articulated key features of scientific practices and made explicit how to develop and refine practical work to focus on specific skills. The authors highlight the need to develop strategies to focus on these specific practices and the need to develop compatible assessment approaches. As examples of how these skills might be assessed in practice, they draw on a separate project in which they refined a set of assessment tasks that situated inquiry skills within engaging contexts and procedures for teachers to make judgments about student performance levels. Tytler and White reflect on their experiences of working with teachers with examples of tasks and rubric development. In a more extensive analysis, Fang and Hsu explore current assessments used for STEM learning through a review of research on STEM (Chap. 12). The authors address the problems that emerged in the review by proposing a multilevel-multifaceted STEM assessment framework to support the design and development of useful assessments for STEM learning. For example, they identify STEM-related competencies as problem-solving abilities and computational thinking skills.

The book raises questions about the role of school culture in promoting STEM teaching and learning. Curriculum and assessment frameworks can be designed to facilitate integration of STEM in educational contexts. Teaching can be enhanced through professional development. The role of domain knowledge can be considered. The particular learning needs of the students can be used to inform teaching. However, without a change in school cultures, it is questionable to what extent the implementation of STEM education will be effective at the school level. In Chap. 10, Widjaja, Loong, Hubber, and Aranda report a design-based research methodology in a case study of two teachers from one secondary school that was implementing an interdisciplinary approach in STEM. The findings of their research indicated increased student engagement and enjoyment from the use of real-world problems and apparent student autonomy and ownership of the project. The teachers thought more deeply about how to better integrate interdisciplinary skills and knowledge in their teaching, and they experienced a greater sense of satisfaction in making learning meaningful for students. The challenges that teachers face include the integration of the learning of two subject disciplines into one interdisciplinary discipline that has real-world applications and the time needed to plan collaboratively. The authors draw attention to how an interdisciplinary approach requires a change in school culture as to the provision in the timetable for interdisciplinary planning and to the enculturation of inquiry-based learning.

Overall, the chapter authors of this book convincingly demonstrate the many opportunities and successes, as well as challenges, in making STEM education a reality in everyday schools. The diversity of national contexts provides insight into how different educational systems are approaching STEM education, including potential integration of the arts as in the case of the Korean curriculum (see Chap. 2). An implicit assumption in the book as well as in the broader community engaged in STEM education is that the integration of science, technology, engineering, and mathematics is justified on the basis of how these disciplines share particular approaches such as critical thinking and problem-solving. The role of domain knowledge is also widely recognized where the particular nuances of concepts and problems in mathematics and science, for instance, are problematized (see Chap. 8). However, the fundamental question remains about which aspects of the STEM disciplines are being considered for educational purposes. For instance, reference to data collection and analysis in STEM concerns particular epistemic practices underpinning the disciplines; reference to the need to foster critical thinking because this is an important aspect of STEM disciplines implicitly refers to the cognitive aspects of STEM as well as scientific ethos surrounding STEM communities (i.e., particular norms about critically evaluating claims and, thus, fostering a sense of skepticism).

13.3 Implications for Future Research

Future efforts in STEM education research and development will benefit from a consideration of not only a fuller articulation of the epistemic and cognitive features of STEM subjects but also the broader social and institutional dimensions. In our recent work, we proposed a framework on the nature of science (NOS; Erduran & Dagher, 2014) that considers NOS as a cognitive-epistemic and social-institutional system (see Table 13.1). In other words, a *systems approach* is used to highlight some high-level categories through which the various dimensions of science can be articulated as part of a bigger system.

The categories can be interpreted in both a domain-specific (e.g., chemistry, biology) as well as domain-general (e.g., science versus non-science) manner. By implication, the framework can be applied to STEM education in order to clarify the various nuances between the disciplinary commonalities and differences between science, technology, engineering, and mathematics. For example, even though all STEM disciplines use observations,

- Can there be variations in the way that engineers and scientists engage in observations?
- What organizations drive the technology world and how do they compare to those where mathematicians are situated?

The categories are comprehensive enough to capture such diverse questions. For example, they relate to the issues of interdisciplinarity and school culture raised in Chap. 10 as well as the issue of epistemic practices of the disciplines raised in

Table 13.1 Cognitive-epistemic and social-institutional dimensions of nature of science (from Erduran & Dagher, 2014)

Cognitive-epistemic system dimensions	
Aims and values	The scientific enterprise is underpinned by adherence to a set of values that guide scientific practices. These aims and values are often implicit and they may include accuracy, objectivity, consistency, skepticism, rationality, simplicity, empirical adequacy, prediction, testability, novelty, fruitfulness, commitment to logic, viability, and explanatory power
Scientific practices	The scientific enterprise encompasses a wide range of cognitive, epistemic, and discursive practices. Scientific practices such as observation, classification, and experimentation utilize a variety of methods to gather observational, historical, or experimental data. Cognitive practices such as explaining, modeling, and predicting are closely linked to discursive practices involving argumentation and reasoning
Methods and methodological rules	Scientists engage in disciplined inquiry by utilizing a variety of observational, investigative, and analytical methods to generate reliable evidence and construct theories, laws, and models in a given science discipline, which are guided by particular methodological rules. Scientific methods are revisionary in nature, with different methods producing different forms of evidence, leading to clearer understandings and more coherent explanations of scientific phenomena
Scientific knowledge	Theories, laws, and models (TLM) are interrelated products of the scientific enterprise that generate and/or validate scientific knowledge and provide logical and consistent explanations to develop scientific understanding. Scientific knowledge is holistic and relational, and TLM are conceptualized as a coherent network, not as discrete and disconnected fragments of knowledge
Social-institutional system dimensions	
Professional activities	Scientists engage in a number of professional activities to enable them to communicate their research, including conference attendance and presentation, writing manuscripts for peer-reviewed journals, reviewing papers, developing grant proposals, and securing funding
Scientific ethos	Scientists are expected to abide by a set of norms both within their own work and during their interactions with colleagues and scientists from other institutions. These norms may include organized skepticism, universalism, communalism and disinterestedness, freedom and openness, intellectual honesty, respect for research subjects, and respect for the environment

(continued)

Table 13.1 (continued)

Cognitive-epistemic system dimensions	
Social certification and dissemination	By presenting their work at conferences and writing manuscripts for peer-reviewed journals, scientists' work is reviewed and critically evaluated by their peers. This form of social quality control aids in the validation of new scientific knowledge by the broader scientific community
Social values of science	The scientific enterprise embodies various social values including social utility, respecting the environment, freedom, decentralizing power, honesty, addressing human needs, and equality of intellectual authority
Social organizations and interactions	Science is socially organized in various institutions including universities and research centers. The nature of social interactions among members of a research team working on different projects is governed by an organizational hierarchy. In a wider organizational context, the institute of science has been linked to industry and the defense force
Political power structures	The scientific enterprise operates within a political environment that imposes its own values and interests. Science is not universal, and the outcomes of science are not always beneficial for individuals, groups, communities, or cultures
Financial systems	The scientific enterprise is mediated by economic factors. Scientists require funding in order to carry out their work, and state and national level governing bodies provide significant levels of funding to universities and research centers. As such, these organizations have an influence on the types of scientific research funded and ultimately conducted

Chap. 11. They can act as tools to organize, analyze, or scrutinize what aspect of STEM is included or excluded in STEM education. They can point to what might be missing in existing STEM initiatives including, for instance, the power dynamics that might govern how STEM professionals engage with each other or with other stakeholders such as the industry and political organizations.

The international curriculum standards landscape is now ripe for the need for such articulation. Consider, for instance, the following passage from the influential *Next Generation Science Standards* in the USA:

> The goal of engineering design is to find a systematic solution to problems that is based on scientific knowledge and models of the material world. Each proposed solution results from a process of balancing competing criteria of desired functions, technical feasibility, cost, safety, aesthetics, and compliance with legal requirements. The optimal choice depends on how well the proposed solutions meet criteria and constraints. (National Research Council, 2013, p. 409)

Such statements call for consideration of the epistemic aims of science and engineering including modeling (i.e., aims and values category from Table 13.1) as well as the broader social features such as costs associated with design solutions (i.e., financial systems category). Furthermore, the disciplinary nuances in what counts as methods in science versus engineering, as well as how knowledge gets validated through social certification in technology versus mathematics, can be researched. In terms of educational development, STEM education will benefit from a systemic evaluation of how STEM curricula and resources as well as instructional strategies and assessment frameworks indeed encompass the nature of STEM in a way that is inclusive of its various dimensions and not only the epistemic and cognitive dimensions.

Engagement in the broader social and institutional aspects of STEM further raises questions about the stakeholders (i.e., teachers, students, curriculum developers, assessment designers) engaged in the design, implementation, and evaluation of STEM initiatives. In other words,

- Who is being engaged in STEM education?
- What is their disciplinary affiliation?
- What community norms are they using to be engaged in STEM education?

The design of teachers' professional development programs, for instance, will benefit from a broad but nuanced take on the disciplinary similarities as well as differences in the STEM subjects. After all, many teachers will themselves have been trained to be part of a professional community that calls for disciplinary specificity. A science teacher is not only someone who is preoccupied with the teaching of the subject knowledge of science; he or she is also someone who is a member of a particular professional community by training. Indeed, even within the particular domain of science, one would encounter the issue of identity among teachers, identifying themselves as a physicist, chemist, biologist, or earth scientist (e.g., Vareals, House, & Wenzel, 2005). These aspects are not simply interesting contextual features—they are paramount to how teachers will understand and perceive STEM and engage in its teaching.

Much research and development work remains to be done in STEM education. A systemic approach to understanding what counts as STEM for STEM education is likely not only to clarify the underpinning theoretical constructs of STEM education but also to open up new avenues for research and development. For example, the approach raises questions such as:

- What are the institutional contexts of STEM professionals?
- Which professional communities do engineers and mathematicians operate in and how?
- What are the implications of their practices for educational practice?
- Are their differences between how scientists and artificial intelligence experts conceptualize models?
- How can we ensure that future STEM professionals can be enculturated into the community norms of their disciplines effectively?

This book provides a robust account of the contemporary space that educators occupy in dealing with related but yet unexplored questions. It provides a research-informed and evidence-based approach; as such, it is a useful resource on which to base future efforts.

References

Erduran, S., & Dagher, Z. R. (2014). *Reconceptualizing the nature of science for science education: Scientific knowledge, practices and other family categories*. Dordrecht, The Netherlands: Springer.

National Research Council. (2013). *Next generation science standards: For states, by states*. Washington, DC: National Academies Press. https://doi.org/10.17226/18290.

Vareals, M., House, R., & Wenzel, S. (2005). Beginning teachers immersed into science: Scientist and science teacher identities. *Science Education, 89*, 402–516.